# Thermoforming:

## Improving Process Performance

# Thermoforming:

## Improving Process Performance

Stanley R. Rosen

Society of Manufacturing Engineers

Plastics Molders & Manufacturers Association of SME

Dearborn, Michigan

987654321



Library of Congress Catalog Card Number: 2002108065

International Standard Book Number: 0-87263-582-1

Additional copies may be obtained by contacting:
Society of Manufacturing Engineers
Customer Service
One SME Drive, P.O. Box 930
Dearborn, Michigan 48121
1-800-733-4763
www.sme.org

SME staff who participated in producing this book:

Philip Mitchell, Editor
Rosemary Csizmadia, Production Supervisor
Kathye Quirk, Graphic Designer/Cover Design
Frances Kania, Production Assistant
Jon Newberg, Production Editor

Printed in the United States of America

# Dedication

To **E. Bowman (Bo) Stratton,** a thermoforming pioneer who is honored for his efforts to build the industry.

# Table of Contents

## 3 Design of Thermoformed Products



## 4 Thermoforming Machines



## 5 Trim Presses

## 9 Mold Cost Estimating



## 10 Knife-like Trim Dies



## 11 Off-line Punch-and-die Trimming

# Preface

The thermoforming industry is naturally divided into two divisions utilizing either thin- or heavy-gage (thickness) plastic sheet. The type of machinery required, the size of production runs, and markets served are different for these two branches of the industry.

*Thin-gage thermoforming* processes plastic sheet of approximately 0.005–0.100 in. (0.13–2.54 mm) thick, which is supplied in roll form or, in special circumstances, extruded directly into the thermoformer. Thin-gage production thermoforming machines can not accept individual sheets and are fed from a continuous roll of plastic sheet. This process is referred to as *roll-fed thermoforming*. *Sheet-fed thermoforming* utilizes heavy-gage sheets cut into individual units, because extrusion equipment limitations prevent the sheets from being coiled into rolls as it is difficult to then unwind the plastic sheet at the thermoformer.

Roll-fed thermoforming machines often form multi-cavity shots at speeds of 3–20 cycles per minute. This process is useful for production runs of 5,000 to multi-million individual parts, such as packaging components, blister packs, and single-use food containers. Sheet-fed thermoforming equipment can form thin-gage parts, but it results in low production rates and high labor content and makes this process uneconomical. Heavy-gage thermoformers produce shots at a cycle speed measured in minutes, rather than in seconds. They are often used for work covering large plastic areas. These large formed parts are used for scientific equipment enclosures, pick-up-bed truck liners, and other industrial products that require high physical strength and durability.

The thermoforming process itself remains much the same for thick or thin sheets, but the molds and trimming techniques can differ. Roll-fed, thin-sheet, thermoforming molds must be temperature-controlled in a sophisticated manner to achieve the high-speed cycling required for efficiency. Sheet-fed molds tend to be less efficiently cooled, perhaps because of smaller production runs and the additional mold cost involved. The trimming of individual thin-gage parts is processed in dies; heavy-gage parts are cut out with routers and saws.

The design principles used for thermoformed parts and molds can be equally applied to either roll or sheet forming. Both industries use vacuum- or pressure-forming techniques on both male and female molds. Achieving nearly uniform wall thickness may require the use of plugs or mechanical assists. Because of the diverse types of machines used, some thermoforming practices may vary to take advantage of equipment capabilities.

Although most details of this text are about roll-fed thermoforming, the principles and some practices apply equally to sheet-fed thermoforming. Physical laws governing heat transfer, mold spacing, location of thick or thin part-wall thickness, mold building, and techniques are universal for any style of thermoforming.

For further information, contact the author:

E-mail: plastimach@plastimach.com
Web: www.plastimach.com

# Abbreviations

## A

| | |
|---|---|
| ABS | acrylonitrile-butadiene-styrene |
| AISI | American Iron and Steel Institute |
| APET | amorphous polyethylene terephthalate |
| ASTM | American Society for Testing and Materials |

## C

| | |
|---|---|
| CAD | computer-aided design |
| CAD/CAM | computer-aided design/computer-aided manufacturing |
| CAE | computer-aided engineering |
| CG | center of gravity |
| CNC | computer numerical control |
| CP | center of pressure |
| CPET | crystalline polyethylene terephthalate |

## D

| | |
|---|---|
| DIN | Deutsches Institut für Normung |

## E

| | |
|---|---|
| EDM | electrical discharge machining |
| ESCR | environmental stress-crack resistance |
| EVA | ethylene-vinyl acetate |
| EVOH | ethylene-vinyl alcohol |

## H

| | |
|---|---|
| HDPE | high-density polyethylene |
| HDT | heat distortion temperature |
| HIPS | high-impact polystyrene |

## I

| | |
|---|---|
| ICP | impact copolymer |
| ISO | International Organization for Standardization |

## L

| | |
|---|---|
| LDPE | low-density polyethylene |
| LLDPE | linear low-density polyethylene |

## M

| | |
|---|---|
| MDPE | medium-density polyethylene |
| MEK | methyl ethyl ketone |
| MFR | melt-flow rate |
| MVTR | moisture vapor transmission rate |
| MW | molecular weight |
| MWD | molecular weight distribution |

## O

| | |
|---|---|
| OEM | original equipment manufacturer |
| OPS | oriented polystyrene |
| OSHA | Occupational Safety and Health Administration |

## P

| | |
|---|---|
| PA | polyamide (nylon) |
| PC | polycarbonate |
| PE | polyethylene |
| PEEK | polyether-ether-ketone |
| PET | polyethylene terephthalate (polyester) |
| PETG | polyethylene terephthalate glycol |
| PP | polypropylene |

| | |
|---|---|
| PS | polystyrene |
| PSU | polysulfone |
| PVA | polyvinyl alcohol |
| PVAC | polyvinyl acetate |
| PVC | polyvinyl chloride |

## Q

| | |
|---|---|
| QA | quality assurance |
| QC | quality control |
| QMC | quick mold changing |

## R

| | |
|---|---|
| $R_C$ | Rockwell C |
| RCP | random copolymer |
| RF | radio frequency |
| RPET | recycled polyethylene terephthalate |
| RPVC | rigid polyvinyl chloride |
| rpm | revolutions per minute |

## S

| | |
|---|---|
| SAN | styrene-acrylonitrile |
| SCR | silicon-controlled rectifier |
| sec | second (time) |
| SPPF | solid-phase pressure forming |
| STAT | sheet thinning analysis (for) thermoforming |

## T

| | |
|---|---|
| TFE | tetrafluoroethylene |
| $T_g$ | glass-transition temperature |

## U

| | |
|---|---|
| UHMW-PE | ultra-high molecular-weight polyethylene |
| UL | Underwriters' Laboratories |
| UV | ultraviolet |

## Reference

Mitchell, Philip, ed. 1998. *Tool and Manufacturing Engineers Handbook*, Fourth Edition, *Volume 9: Material and Part Handling in Manufacturing*. Dearborn, MI: Society of Manufacturing Engineers.

# 1

# The Roll-fed Thermoforming Process

## Introduction

*Thermoforming* is a term of recent origin, perhaps coming into common use in the early 1980s. Before that time, sellers of thermoformed plastic products referred to their industry as either vacuum- or pressure-forming operations. It made sense to combine the definitions of these processes since the only difference between the two is the degree of air pressure applied to form heated plastic. Thermoformed product buyers do not care which process is used as long as finished products are of high quality, delivered on time, and sold at a fair price.

## Heating the Plastic

The thermoforming process requires that a plastic sheet be heated uniformly to a temperature that makes it soft enough to process. The best temperature for thermoforming is often arrived at empirically, since the true measurement of sheet temperature can be difficult. If the sheet temperature rises above the upper thermoforming limit, the plastic material may degrade or sag, causing sheet thinning and a possible loss of properties (see Figure 1-1). When the sheet temperature is below its lower forming limit, the material may cold form—inducing internal stresses that later can precipitate part failure.

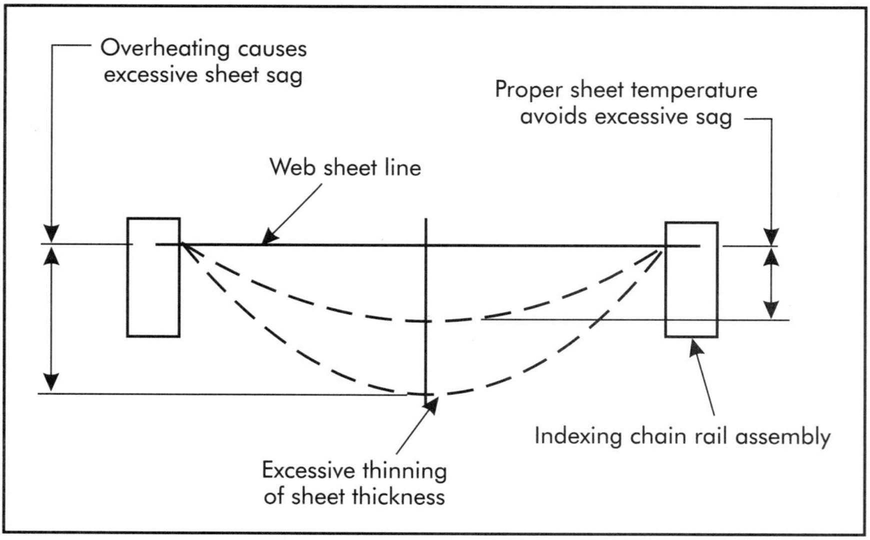


*Figure 1-1. Sheet thinning and a possible loss of properties.*

Mold design cannot compensate for thermoforming at an incorrect sheet temperature. A visit to a thermoforming shop will reveal the ongoing struggle to maintain the correct forming temperature range. Thermoformed plastic parts must start with a consistent, uniform sheet temperature or production will yield a variable-quality product.

## Thermoforming a Continuous Web

Roll-fed thermoforming is a continuous process. When a sheet from one roll is exhausted, the next roll web is attached to allow continuous production. Rolls of plastic sheet vary in width from 10–60 in. (254–1,524 mm) with thickness ranging from 0.0075–0.1000 in. (0.190–2.540 mm) or greater and weighing from 50–2,000 lb (22.7–907 kg) wound on cardboard or metal cores (see Figure 1-2).

When only one family of plastic resin is used for a large quantity of products—for example cups and lids—an in-line extruder can be installed to directly process the sheet continuously into

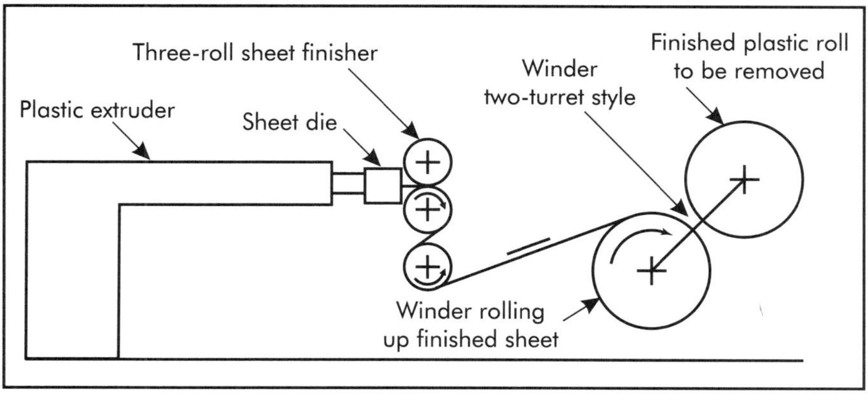


*Figure 1-2. Rolls of extruded plastic sheet being wound on cardboard or metal cores (Mitchell 1998).*

the thermoformer (see Figure 1-3). The roll-fed thermoforming machine will accept a plastic web equally from rolls or an extruder. Proponents of direct extrusion claim they achieve a more uniformly heated sheet because the center core of the sheet's cross-section tends to remain hot when entering into the thermoformer oven.

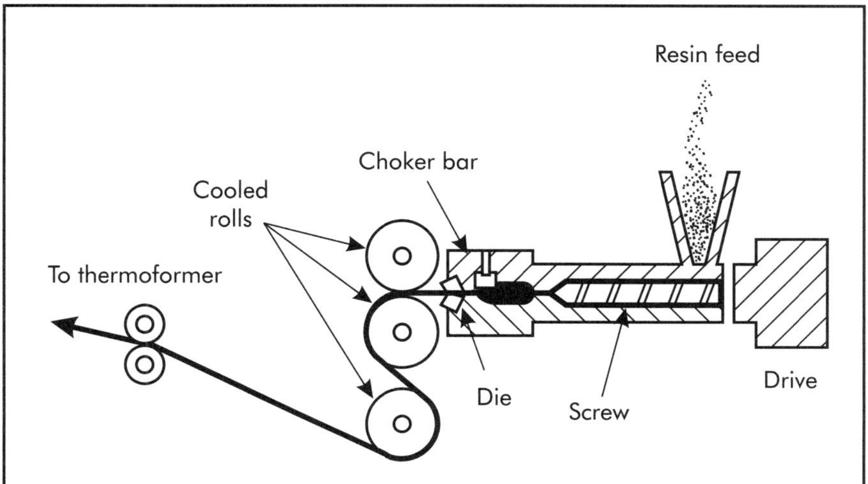


*Figure 1-3. An in-line extruder installed to directly process the sheet continuously into the thermoformer (Mitchell 1998).*

## Machine Components

The continuous roll-fed thermoforming system consists of the following components:

- A roll stand that accepts and unwinds a plastic roll (see Figure 1-4) to serve the thermoforming machine. This device generally has space for two rolls of plastic to reduce downtime when the first roll is emptied.
- Ovens with radiant heating elements are capable of heating a plastic web from one side or both sheet faces.
- An indexing chain system transports the plastic web off of the roll stand, through the ovens and thermoforming press, to a trim station and packing area. The length of the index stroke is the same length as that of the mold, plus a clearance factor. It is critical that the indexed web be consistent and accurate from point to point so that the trim press die can more easily locate the formed part precisely for accurate trimming.
- Thermoforming presses that contain the mold are placed immediately next to the oven exit so the heated sheet does not have an opportunity to cool before being indexed into the mold (see Figure 1-5). Most, but not all, U.S.-built thermoformers have both an upper and lower moving platen in the press section; and many presses have mechanical devices incorporated to operate mold ejectors. The forming press area has piped utility outlets, which provide for vacuum and pressure forming, mold temperature control of liquids, and part ejection air. Some projects may require vacuum and pressure used in tandem to form plastic components.
- On-line trim presses are built on the machine frame and are fed by the indexing system of the thermoformer to register a whole shot relative to the die.
- Off-line trim presses are stationed to pick up the free web after exiting the thermoformer, and incorporate an independent indexer to align one or more formed rows of parts to the die.
- Auxiliary devices may be present, such as stacking or collating systems to organize trimmed parts for packing and counting once the parts exit the on-line trim press.

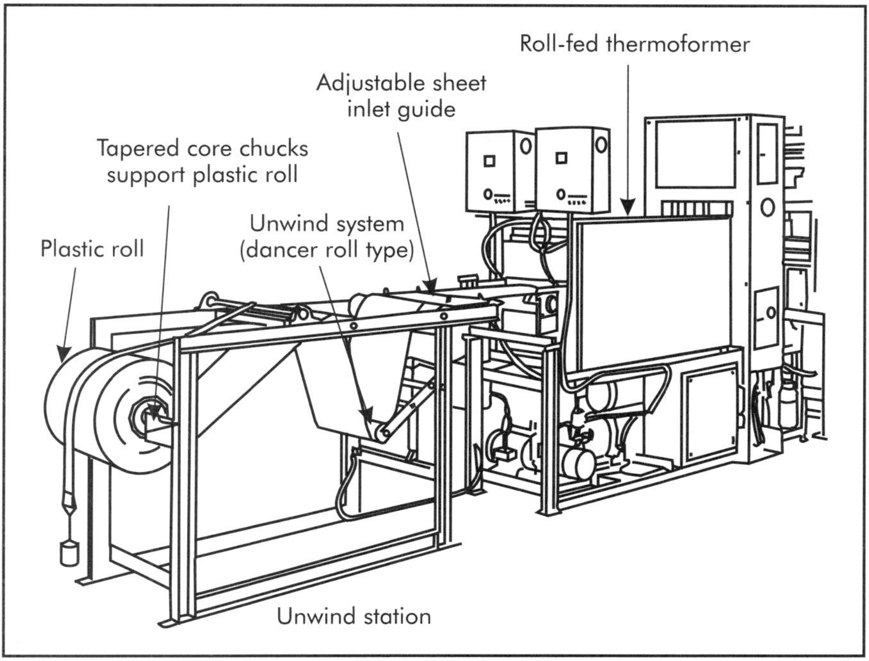


*Figure 1-4. A roll stand that can accept and unwind a plastic roll (Mitchell 1998).*

The decision on which cutting system is selected for trimming involves many factors—initial cost, delivery time, quality of cut edge, available trimming equipment, and the skill of thermoforming setup persons.

## Vacuum-forming Process

Vacuum-forming tooling requires a mold connected to a vacuum supply and a system to seal the edges of the mold securely against heated plastic (see Figure 1-6). The seal-off system is necessary for the vacuum to build up rapidly between the mold and heated plastic without leakage. Vacuum leaks can cause trapped air between the mold and hot sheet to cool the plastic sufficiently and prevent it from coming in contact with the mold surface. If the hot plastic becomes chilled before it is fully thermoformed, it may

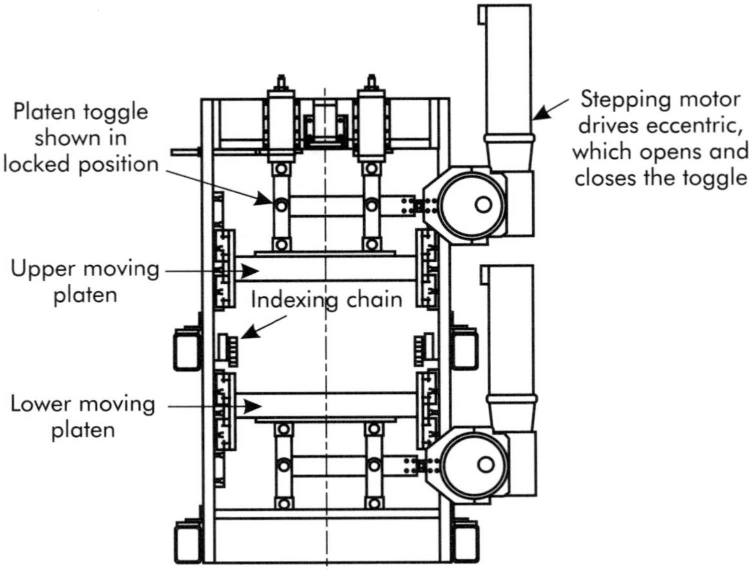


*Figure 1-5. Thermoforming presses to contain the mold assembly (Mitchell 1998).*

result in a component that is incompletely formed and of poor quality. Unfortunately, a small or medium vacuum leak is difficult to detect and trace. It displays little evidence of its location. A consequence may be a request to the mold maker to rework the vacuum path to increase airflow. However, this modification may not solve the problem. A properly designed seal-off method and the intervention of a skilled thermoforming operator tends to minimize this difficulty.

## Pressure-forming Process

Pressure forming requires a pressure box to enclose the mold and create a pressure-tight thermoforming area, which allows thermoforming using pressures above 14.7 psi or one atmosphere or 101 kPa (see Figure 1-7). There is no problem in detecting pressure-forming leaks; pressurized air will advertise the leak with a mighty roar.

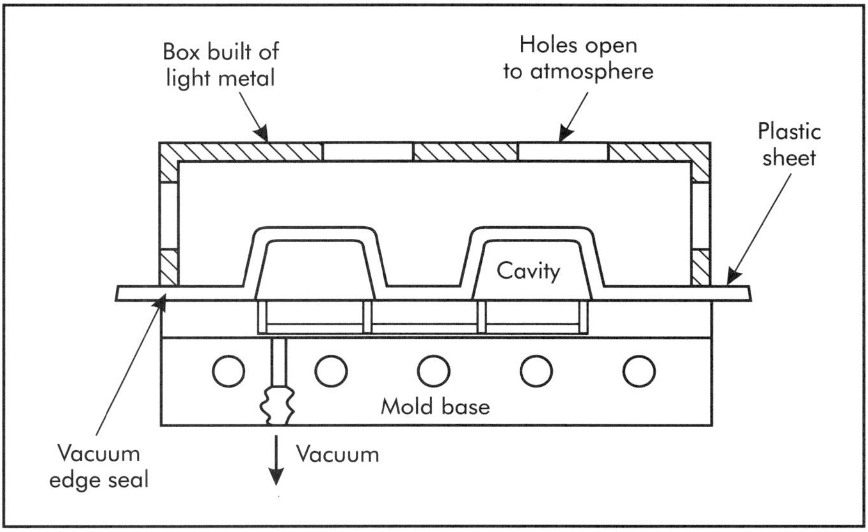


*Figure 1-6. Tooling for vacuum forming (Mitchell 1998).*

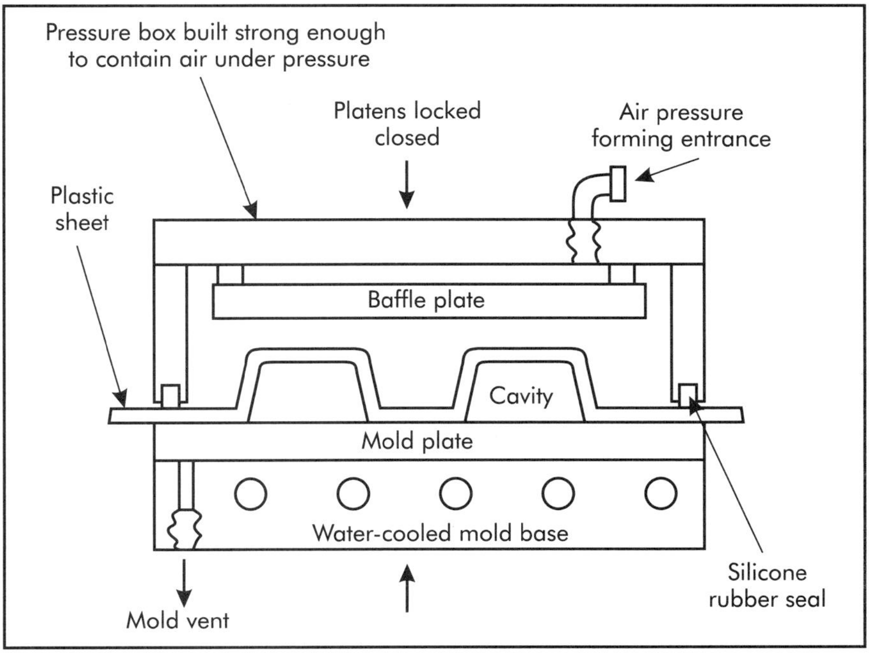


*Figure 1-7. A pressure box to enclose the mold and create a pressure-tight thermoforming area (Mitchell 1998).*

## Comparison of Processes

Both vacuum- and pressure-forming tooling often require plugs or assists to help pre-stretch heated plastic to achieve a uniform part-wall thickness before activating the forming sequence. The tool may need special devices to eject parts that tend to hang up in a mold, such as cavities with undercuts. The mold-cavity design for a male or female cavity is exactly the same for pressure and vacuum forming.

A uniformly heated sheet of most commonly specified resins produces equally acceptable parts on a vacuum or pressure former. Unfortunately, all sheets are not uniformly heated; nor is every machine operator equal in skill and knowledge. The higher force of pressure forming permits the temperature window to be wider so that acceptable parts are produced more consistently over the lower end of the thermoforming range. Overall scrap rate is reduced by a considerable factor, thereby increasing profitability. Because pressure forming forces plastic into closer contact with the mold, heat transfer from the plastic to the metal mold is more efficient, thereby increasing cycle speed and resulting in more parts produced per shift.

Roll-fed vacuum forming is appealing only for those short runs that cannot amortize the purchase price of a pressure box and a modern temperature-controlled mold.

## Trimming

A continuous roll-fed thermoformer can produce a multi-cavity shot for each machine cycle. Individual components must be trimmed from the shot to provide a usable thermoformed product. Thermoforming and trimming are equally important in the manufacture of finished components and each process requires its own high degree of technical knowledge and skill.

Two methods of trimming are employed to cut out roll-fed thermoformed plastic parts. The earliest cutting technique adapted for thermoforming came from the paper industry, where knife-like, steel-rule dies are used to cut out card boxes, printed material, jigsaw puzzles, etc. (see Figure 1-8). When a high volume of thermo-

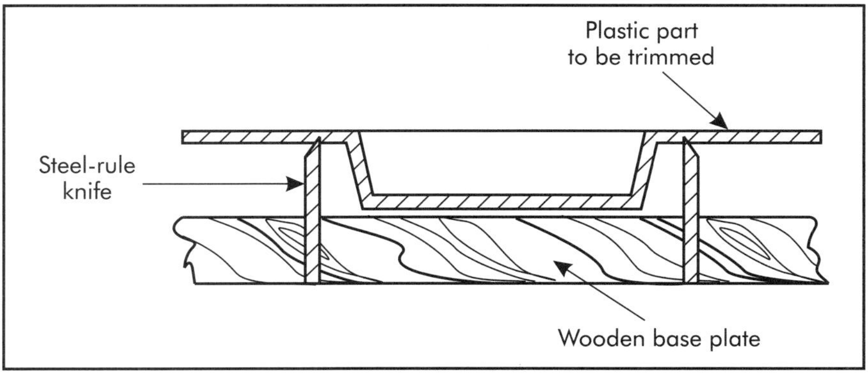


*Figure 1-8. The earliest cutting technique adapted for thermoforming is a steel-rule die (Mitchell 1998).*

formed parts is required, the most efficient trimming system is the punch and die, a method taken from the metal stamping industry (see Figure 1-9). Equipment is available that can form and trim components within the mold using several techniques.

Knife-like dies function in the same way that a pocketknife cutting through wood acts when its sharp thin edge is forced into a material of lesser hardness. The duller the cutting edge becomes, the greater the force required and the resultant trim perimeter quality suffers. Each plastic resin has specific trimming characteristics—some soft-yet-tough or hard-but-brittle and some that

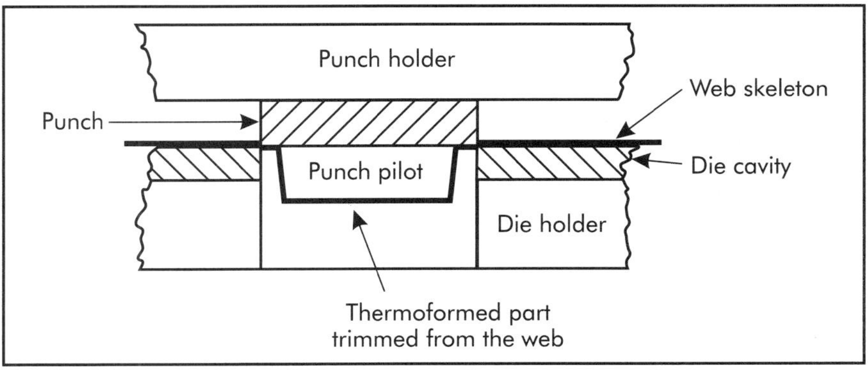


*Figure 1-9. A punch and die is the most efficient trimming system when a high volume of thermoformed parts is required.*

are very abrasive to sharpened blades. Knife-like dies come in many types of bevels, hardness levels, degrees of bendability, and levels of thickness. These factors must be weighed when specifying these dies.

A knife-like tool system consists of a die, striker plate (which the die impinges on), and a buildup to support the die and striker plate against deflecting under load (see Figure 1-10). The striker plate surface can be softer or harder than the die, depending on the skill and patience employed in setup of the trim press. The best situation is one in which the die *kisses* but does not penetrate the striker plate. A die allowed to penetrate into the striker plate can dull on the first attempt at cutting if it is softer than the plate. If a softer plate than the die is used, the knife edge can remain sharp, but it creates a groove in the striker plate that eventually becomes wider. The plastic then tends to bend and break in the groove rather than cut through cleanly.

## Punch and Die

Punch-and-die tooling includes some of the following components:

- Punch-and-die modules consist of the punch with a male configuration that enters with a slight sidewall clearance into the female die (see Figure 1-11). Because of the high cost of constructing a multi-cavity die, only one or two rows are trimmed with each stroke of the trim press.
- A die set specified with four posts and bushings is required to guide the moving and stationary plates and align the punch-and-die module components to each other.
- Pilots built on the punch help center the formed shots to the punch and die module.
- Knockouts are included, where applicable, to push trimmed parts free of the punch and into the die cavity.
- Spring-loaded stripper plates clamp the sheet to be trimmed firmly against the die plate for a clean cut. After the part is trimmed, the stripper plate removes the sheet skeleton from the punches.
- Chutes stack the trimmed parts in uniform order for ease in safely removing and packing them.

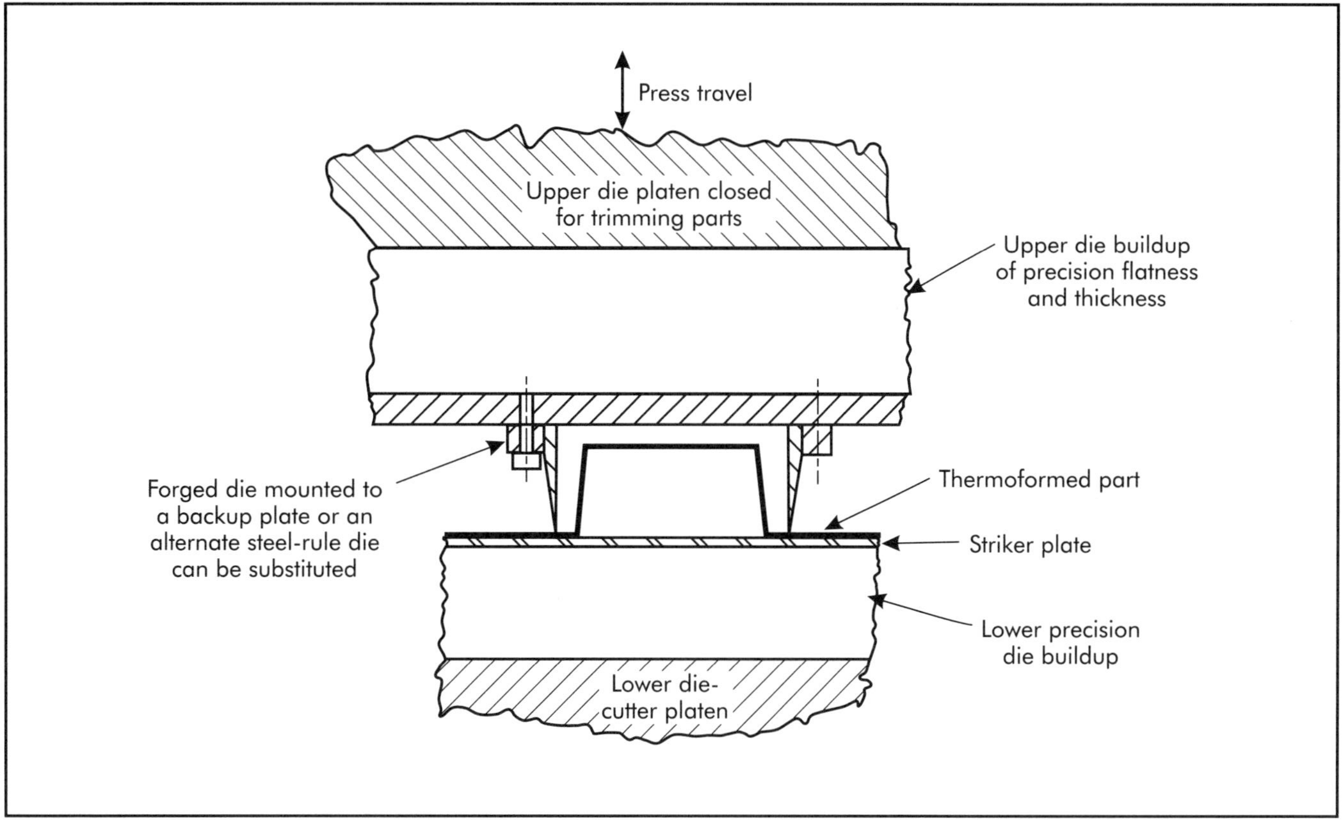


*Figure 1-10. A buildup to support both the die and striker plates to prevent deflecting under a load.*

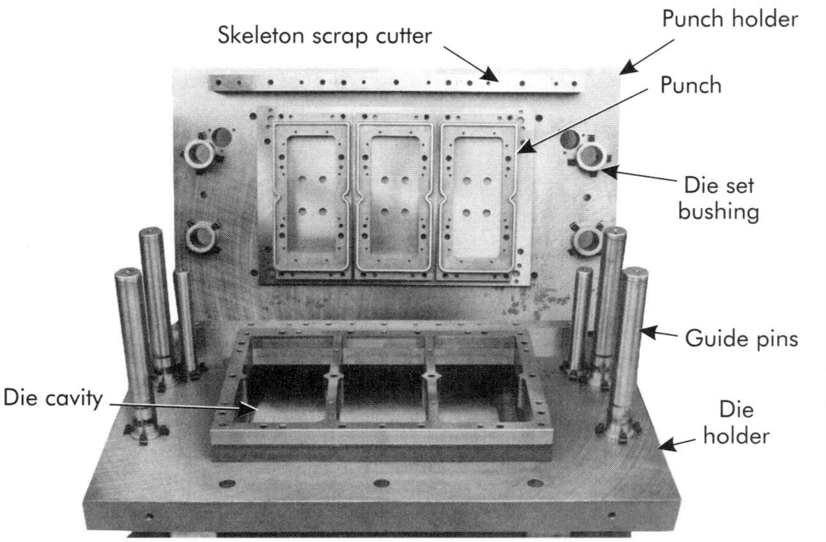


*Figure 1-11. A punch and die assembly. (Courtesy Ontario Die Ltd.)*

## Sheet Indexing

Accuracy for a plastic sheet that is indexed through a thermoforming machine affects the performance of the trimming operation. Trimming on-line with the trim press mounted on the thermoforming machine frame requires that shots be consistently indexed from the forming area to the trim press (see Figure 1-12). Small errors of location can be corrected by building locating pilots into the die. Knife-like dies can be designed to float in the direction of index within a movement of 0.06–0.19 in. (1.5–4.8 mm) to accommodate more erratic indexing.

## Resin Shrinkage and Trim Registration

More difficult registration problems can occur from selecting an incorrect resin shrinkage coefficient for a large multi-cavity part produced on an on-line thermoforming machine. The resin shrinkage coefficient used for design of molds and dies applies

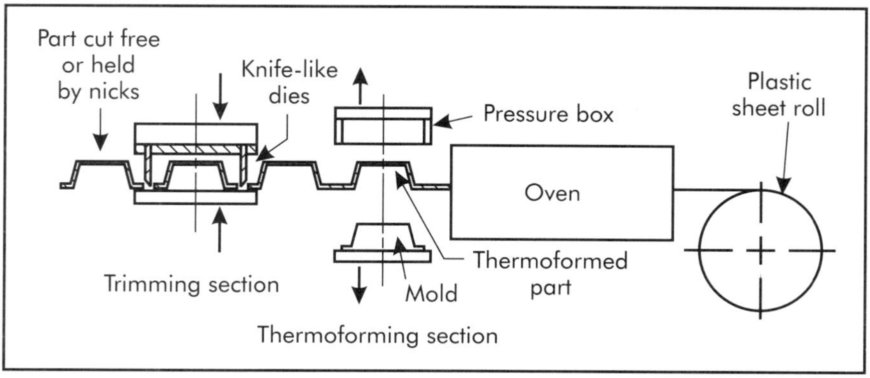


*Figure 1-12. Trimming on-line with the trim press mounted on the thermoforming machine frame (Mitchell 1998).*

only to cold sheets. Often the thermoformed sheet is quite hot on exiting the mold and it may remain hot when it moves into the die. The cold-shrinkage coefficient may be specified at, for example, 0.005 in./in. (0.005 mm/mm). Yet, on the on-line machine, true shrinkage may be 0.002 in./in. (0.002 mm/mm) at the trim station. Miscalculating true shrinkage by 0.003 in./in. (0.003 mm/mm) over a 30-in. (762-mm) part width dimension would multiply to be an error of 0.090 in. (2.29 mm). There is no way to correct errors of this scale unless the mold cavity is designed to be movable or a new die is constructed. Fortunately, this type of error is not common on punch and dies installed on self-indexing off-line trim presses that pick up the shot a minute or two after being formed. By this time, the sheet is usually fully cooled (see Figure 1-13).

## Part Handling

Most thermoformed parts are shipped in neat stacks rather than in a jumble of loose pieces. Stacking requires a method to allow parts to easily de-nest one from the other and not jam together during shipping. The part fabricator is interested in specifying a minimum stack-height allowance between each part so the maximum number of components can be placed in the shipping container.

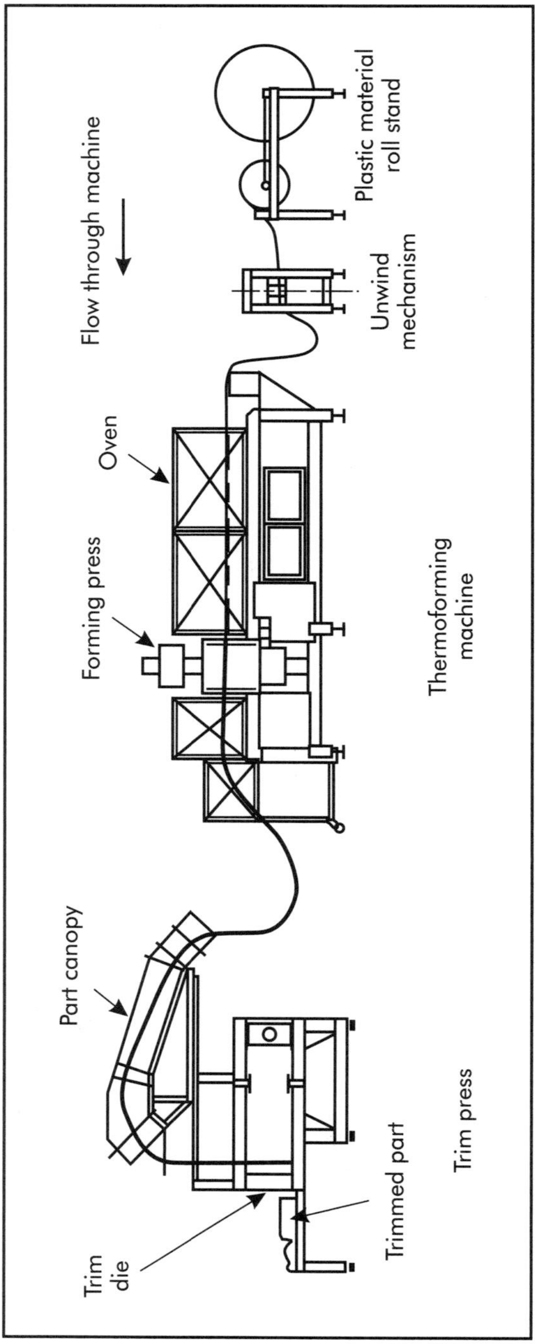


*Figure 1-13. Off-line self-indexing trim press in tandem with a thermoformer (Mitchell 1998).*

The geometry of the parts, draft angle, wall thickness, and type of resin govern de-nesting characteristics, as does skillfully designed, thermoformed de-nesting lugs. A designer can lay out these conditions on paper, but actually stacking finished products formed on a prototype thermoforming machine to the height being shipped makes the final judgment.

On-line thermoformers are able to index cut thermoformed parts out of the trim press because weakened nicks or tabs connect them to the shot (see Figure 1-14). These tabs allow parts to remain in the shot until a light force removes them during packing. Cut parts can be stacked manually after exiting the die press or by an in-line stacking station that automatically strips and loads finished parts into chutes and counts them.

Off-line trim presses, with punch-and-die tooling, trim formed parts out of the web. These parts move through the back of the die and out to a chute where packing takes place.

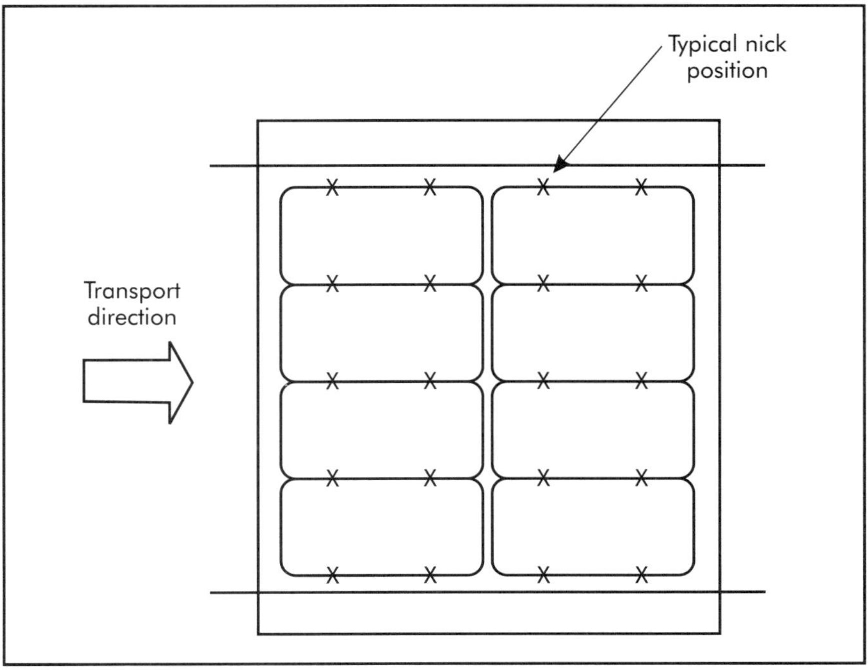


*Figure 1-14. Holding trimmed parts in a web using "nicks," when being indexed out of the steel-rule press.*

## Secondary Processes

Additional secondary processes that can be run on an on-line thermoformer (see Figure 1-15) are as follows:

- Certain thermoformed products are designed with cardboard or plastic sheet backing permanently laminated by adhesive or heat sealing. Many of these common products—such as cosmetic packages and insect traps—need the added strength provided by the backing for the thermoformed container. These specialty products can be completely fabricated on an on-line thermoformer. The trimming and backing board are cemented or heat-sealed and trimmed simultaneously in the trim press using a variation of the knife-like die.
- Hot-stamping equipment can be mounted as an optional station on an on-line thermoformer to decorate formed components. The tool designer may have to develop anvils to back up plastic thermoforms while they are being processed at the hot-stamp station.

The on-line thermoforming system requires tooling for both routine and specialty processes. A designer should think through the complete production process and foresee all tooling needs. Experience indicates that thermoformed parts' scheduling becomes chaotic as delivery deadlines near. A "quick and dirty" production mentality may take hold if all production tools are not provided in time to start operations.

## Reference

Mitchell, Philip, ed. 1998. *Tool and Manufacturing Engineers Handbook*, Fourth Edition, *Volume 8: Plastic Part Manufacturing*. Dearborn, MI: Society of Manufacturing Engineers.

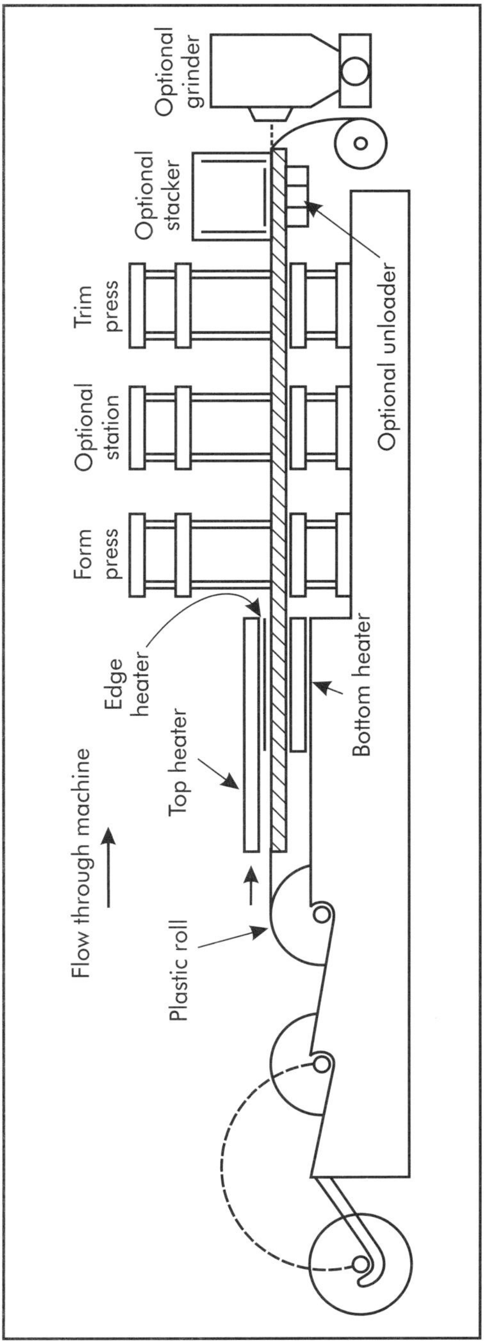


*Figure 1-15. Additional secondary processes that can be processed on an on-line thermoformer (Mitchell 1998).*

# 2

# Properties of Plastic Materials

*Phillip Jacoby, Ph.D., co-author*
*BP Polymers*

The types of plastic resins used in thermoforming are known as *thermoplastics*, and all thermoplastics share certain common characteristics. Thermoplastics will soften when heated, and at some elevated temperature they turn into viscous fluids, much like thick molasses. This allows them to be pumped through a thin die using an extruder to make a thermoformable sheet.

If the structure of a molten plastic is magnified by more than a million-fold, it would consist of long thin chains highly entangled with one another and in rapid motion, almost like a bucket of worms. Each of these chains is a single polymer molecule. If a force or pressure is applied to the melt, such as rotating an extruder screw or moving a plunger, the polymer molecules slide past one another and the fluid undergoes flow. The higher the temperature, the more rapidly the molecules move, and the easier it is to make fluid flow. The fluid thins and drops in viscosity. Viscosity of the polymer melt also depends on how flexible the chains are, and how entangled they are with one another. Further magnification of the structure of each chain shows that it consists of a large number of identical units or links joined together. These units are the *monomers* from which a polymer is formed. The word *polymer* is a combination of the words *poly*, meaning many, and *mer*, meaning a unit. A typical polymer chain may be thousands of monomer units in length and is formed in a process called *polymerization*.

There is another class of polymers known as *thermosets*, which cannot be used in thermoforming. In these materials, the polymer chains are linked together, or cross-linked to form a giant network, much like chicken wire (see Figure 2-1). These cross-links prevent chains from sliding past one another, and thus a thermoset polymer will not form a melt that can be pumped through an extruder. Examples of a cross-linked polymer are the rubber in an automobile tire and hard, brittle materials like phenolic resins used for knobs or handles on cookware.

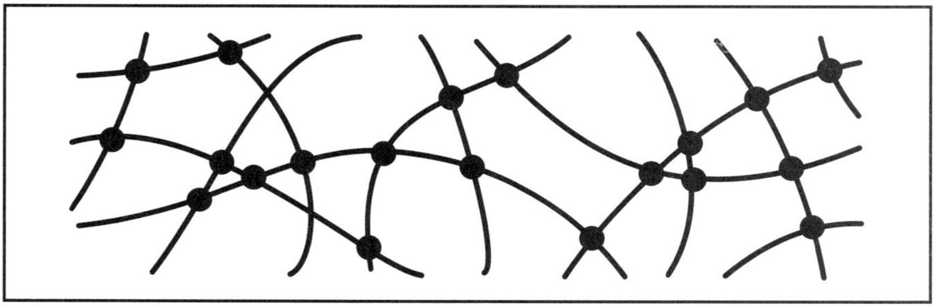

*Figure 2-1. Thermosetting plastics form a three-dimensional random network with rigid bonds. (Courtesy BP Polymers)*

For certain thermoplastics it is possible to have chains that contain long branches or even an occasional link to another chain. An example of two different types of branched polyethylene molecules is illustrated in Figure 2-2. As long as the number of these links is not too high, these materials will form a melt that can be extruded into sheets and thermoformed. This branched structure often leads to a melt with greater strength and rigidity (high melt strength). This allows the sheet to better resist gravity when it is heated, and therefore undergo less sag in the thermoforming oven. This feature can be especially important for improving processing of crystalline thermoplastics.

## Crystalline versus Amorphous Thermoplastics

Solid materials generally occur in one of two forms, either *crystalline* or *amorphous*.

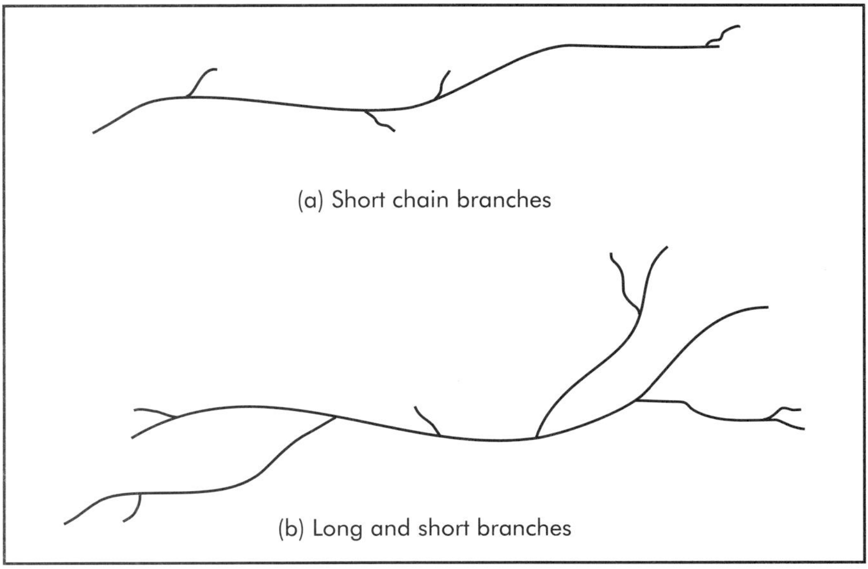


*Figure 2-2. Two different types of branched polyethylene molecules. (Courtesy BP Polymers)*

## Crystalline Material

In a crystalline substance, such as ice, molecules are arranged in a very regular pattern. This ordered arrangement sometimes produces the beautiful geometric crystals of minerals like quartz, or the delicate six-pointed structure of a snowflake. When a crystalline substance is heated, it eventually reaches a temperature where the molecules are moving too rapidly to be held in the crystal, and melting occurs. At this melting point, it rapidly goes from a rigid solid to a fluid melt, and the transition occurs over a very narrow temperature range.

In many materials where the molecules are large and bulky, or chain-like in structure, it is often not possible for molecules to arrange themselves into a regular crystalline structure when the liquid or melt is cooled. In the solid state of these materials, molecules are arranged in a disordered or random fashion, and this material is said to be in the *amorphous* or *glassy* state. Appearance of the molecules can be likened to a bowl of cooked spaghetti (see Figure 2-3).

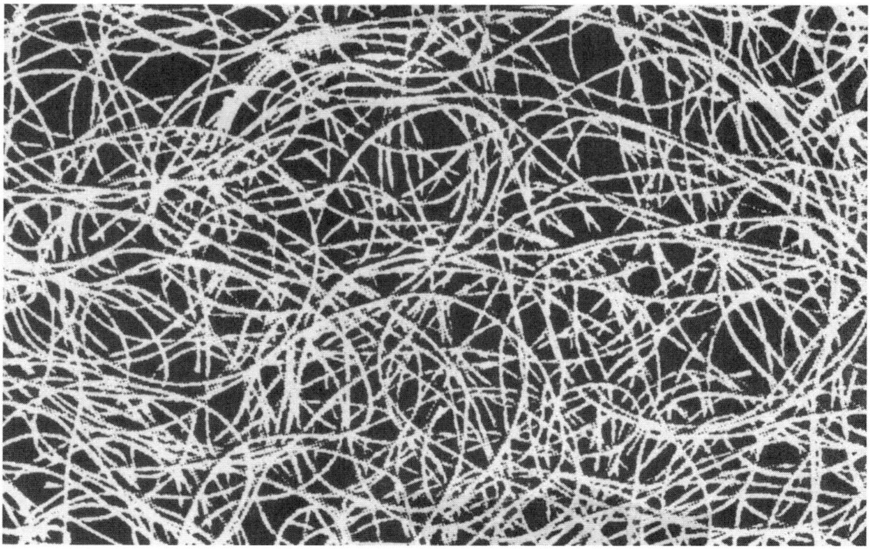

*Figure 2-3. The appearance of the amorphous plastic molecules can be likened to a bowl of cooked spaghetti. (Courtesy BP Polymers)*

## Amorphous Material

Amorphous plastics are often transparent. Typical examples include crystal polystyrene, polymethyl methacrylate (Acrylic™), and polycarbonate (Lexan™). It should be noted that the term *crystal*, which is associated with amorphous polystyrene, refers to the crystal-clear appearance of the plastic, and not crystallinity. Ordinary silica glass is an example of an inorganic amorphous polymer.

When a glassy material (amorphous) is heated, it gradually begins to soften as molecules gain more and more mobility. However, unlike a crystalline material, there is not an abrupt change from a rigid solid to a fluid melt. If a glass rod or tube is held over a flame, it eventually becomes soft enough to bend or blow into a bottle, but unless it is overheated, it will not appear to melt. The temperature where large-scale motion of molecules becomes possible, and significant softening of the material occurs, is known as the *glass-transition temperature*. When an amorphous plastic sheet is heated above its glass-transition temperature in the oven of a thermoformer, it begins to gradually soften and has almost a leather-like appearance over a broad temperature range. This broad softening range gives it a wide processing window to be formed into a final part, without excessive sag in the oven. The abrupt melting of a crystalline plastic leads to a much narrower temperature window over which thermoforming can occur.

## Semicrystalline Plastics

Many types of thermoplastics can crystallize when they cool from the melt (see Figure 2-4). However, because of their entangled chain structure, they cannot achieve the perfect ordering possible for small-molecule materials. When crystallization does occur in polymers, the chains generally line up parallel to one another and fold over to form thin crystals known as *lamellae*. In the melt, these crystals often start growing at the surface of small foreign particles known as *nuclei*. The crystals grow by fanning out in all

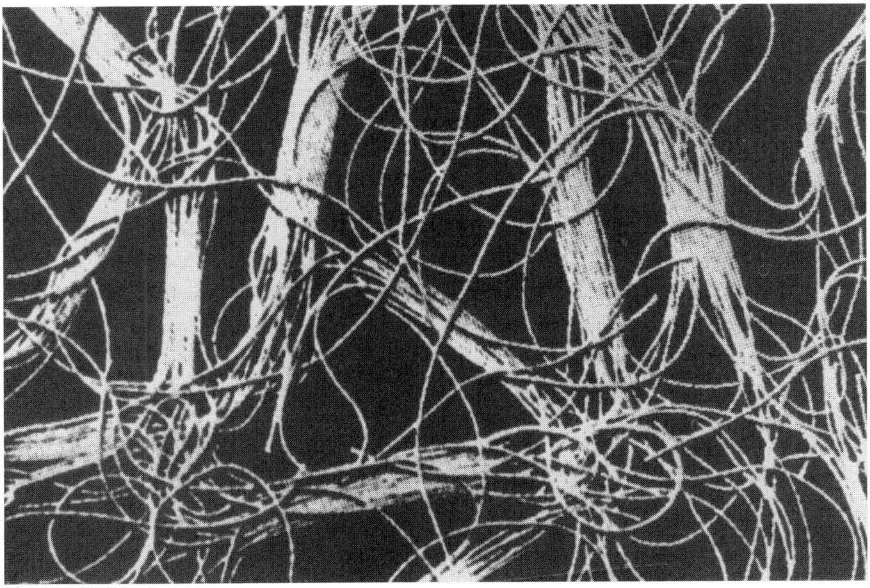

*Figure 2-4. Semicrystalline plastics, such as polyethylene or polypropylene, are translucent or opaque. (Courtesy BP Polymers)*

directions to form spherical bodies called *spherulites*. The folds and entanglement between lamella have a disordered or amorphous-like structure, so that the final solidified polymer is only semicrystalline. Semicrystalline plastics, such as polyethylene or polypropylene, are usually translucent or opaque, since the crystals or spherulites scatter the passage of light through the plastic.

## Crystallinity

The degree or percent of crystallinity depends on the regularity of the polymer chain, and how much opportunity chains have to crystallize before they are frozen in place as the melt cools below the glass-transition temperature. Slow cooling of the melt favors more complete crystallization and a higher level of crystallinity. A rapidly quenched part or sheet will generally have lower crystallinity and better clarity than a slowly cooled material. A high concentration of nuclei also favors higher crystallinity because there are more sites for crystals to form on, and nucleating agents are often added to polymers for this reason. However, nucleated plastics, generally have much improved clarity because they contain smaller spherulites that interfere less with light passing through the polymer.

The simplest example of a semicrystalline polymer is polyethylene (PE). Under ideal conditions, crystallinity levels as high as 80% can be achieved, provided there are not defects in the chain structure. In polyethylene crystals, chains are closer together than they are in amorphous regions. This causes crystalline PE to have a higher density than amorphous PE. A polyethylene with a high level of crystallinity typically has a solid-state density of about 0.95–0.96 grams per cubic centimeter (g/cc), and is known as high-density polyethylene (HDPE).

One way to lower the density or crystallinity of polyethylene is to mix in other co-monomers, such as butene-1 or hexene-1, with the ethylene during polymerization. When these other monomers are inserted along the growing chain, they produce short branches that do not easily fit within the crystal structure. This causes the crystals to be less perfect and less abundant—leading to a lower level of crystallinity and a lower density. When the density of the polyethylene is in the range of 0.90–0.92, it is referred to as low-density polyethylene (LDPE). As density (crystallinity) goes down, the polymer becomes softer and more pliable, and the melting point also decreases. Another way to make LDPE is to polymerize the ethylene in a high-pressure reactor where many long branches are produced along the chain. These branches interfere in the same manner with the development of crystals, although the viscous (flow) behavior of this long-chain branched PE is different from that of the short-chain branched PE. The short-chain branched material is referred to as linear LDPE (or LLDPE) to distinguish it from its long-chain branched counterpart.

## Molecular Weight and Molecular Weight Distribution

The molecular weight of a polymer molecule is a measure of how large the molecule becomes. For example, an ethylene molecule ($C_2H_4$) has an atomic weight of 28 (each carbon is 12 and each hydrogen is 1). If 1,000 ethylene molecules are joined together, the polyethylene molecule formed has a molecular weight of 28,000. Polymerization produces a range of different molecular-weight polymer molecules, so an average molecular weight must be used to describe a molecule. As the average molecular weight of a polymer increases, dramatic changes occur to both melt- and solid-state behavior of the material. A high-molecular-weight polymer has more resistance to flow and is therefore more difficult to extrude into sheet. The toughness or impact strength of a polymer also goes up as molecular weight increases.

Molecular-weight distribution (MWD) also has an effect on physical properties. In this case, MWD refers to the relative amount of very small and very large molecules that make up the plastic. If most of the polymer molecules are similar in size, the polymer has a narrow molecular-weight distribution. On the other hand, if there are large numbers of very small and very large molecules, the polymer has a broad MWD. It is possible for two different samples of a given polymer to have the same average molecular weight, but very different molecular-weight distributions.

## Mechanical Properties of Solid Polymers

This section will discuss the mechanical properties of solid polymers, which include stiffness, modulus, orientation, heat distortion temperature, and toughness.

### Stiffness

A critical property for most thermoformed parts is stiffness or rigidity. A common measure of stiffness is the flexural modulus, which is determined by a three-point bending test on a rectangular beam (see Figure 2-5). The force required to bend or deflect the beam—what we normally think of as the stiffness—depends on the inherent modulus of elasticity of the material and the beam dimensions. Thickness has a very dramatic effect on stiffness, since it appears as a cubic term in the deflection equation (see Equation 2-1).

Maximum deflection can be found by performing a three-point flexure test for modulus measurement:

$$M_d = \frac{PL^3}{4Ebd^3} \qquad (2\text{-}1)$$

where:

$M_d$ = maximum deflection, in. (cm)
$P$ = applied force, lbf (N)
$L$ = test length, in. (cm)
$E$ = modulus of elasticity for sample, psi (kPa)
$b$ = width of sample, in. (cm)
$d^3$ = thickness of sample, in.$^3$ (cm$^3$)

For example:

> If a beam 0.100 in. (2.54 mm) thick requires a force of 50 lb (222 N) to deflect it by 0.005 in. (0.13 mm) at its midpoint, a beam of the same material that is only 10% thicker (0.110 in. or 2.79 mm) requires a bending force of about 65 lb (289 N)—an increase of 30% to produce that same 0.005 in. (0.13 mm) bend. As a general reference, the percentage increase

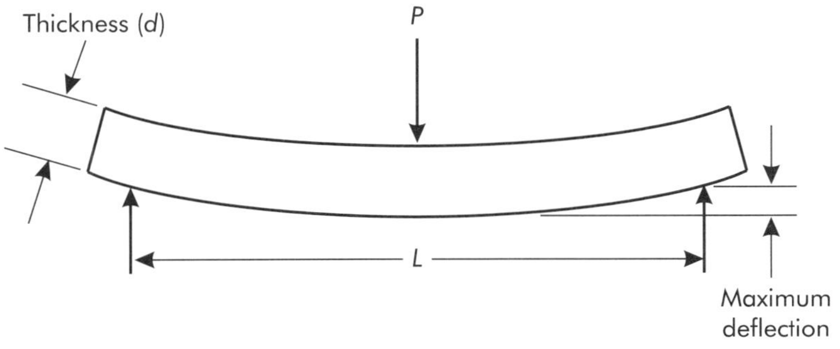


*Figure 2-5. A three-point bending test on a rectangular beam (Mitchell 1998).*

in bending force will be about three times the percentage increase in thickness. This is why having a uniform wall thickness in the formed part is so critical to its performance.

## Modulus

The modulus is a property of the polymer and its processing history. For amorphous polymers, the modulus is directly related to the stiffness of the polymer backbone chain. Very flexible polymers, such as polyethylene, have a low modulus, while polymers with more rigid backbones, such as polystyrene and polycarbonate, have a much higher modulus. For semicrystalline polymers, the modulus of the crystalline regions is higher than that of the amorphous regions, especially at temperatures above the glass-transition temperature of the amorphous phase. The modulus of a semicrystalline polymer will therefore increase as the level of crystallinity increases. In the case of polyethylene, high-density polyethylene (HDPE) is a stiffer material than low-density polyethylene (LDPE). If a nucleating agent is added to a semicrystalline polymer like polypropylene, the final part also has a higher level of crystallinity and a high modulus.

## Orientation

Polymer molecules orient in the direction of applied stress when a polymer is processed by extruding it into a sheet or forming it into a final part. If molecules in a beam are oriented parallel to the long direction of the beam, that beam has a higher modulus than one where there is no preferred orientation of molecules, or where molecules are oriented perpendicular to beam direction. When high levels of orientation occur in the extruded sheet, the sheet tends to distort and shrink when it is reheated in the oven. This is because molecules tend to go back to their unoriented arrangement once they are free to move. If the sheet is thermoformed at fairly cold temperatures, the orientation imparted to molecules in the forming process sets when the part cools in the mold. This results in a final part with greater stiffness and rigidity. For this reason, parts that are made via solid-phase pressure forming (SPPF) are more rigid than those made via the melt-forming process. However, as with the oriented sheet, these highly oriented SPPF parts are more prone to warp and distort if they are reheated (for example, during microwave heating of food) to temperatures close to that of the forming temperature.

The modulus of all plastics decreases as the temperature is raised. This effect influences both final properties of the thermoformed part and processing conditions needed to make these parts. Change of modulus with temperature is quite different for a crystalline versus an amorphous material. This difference is illustrated in Figure 2-6 for polypropylene versus polystyrene.

The glass-transition temperature of polypropylene occurs at about 32° F (0° C). Since polypropylene is a highly crystalline material, and glass transition is a property of the amorphous region, the drop in modulus at this temperature is relatively small. The major drop in the modulus of polypropylene occurs when the melting point is reached (325° F or 163° C) and the crystal phase disappears. The suddenness of this melting transition is the reason why sheet temperature is so critical for successfully thermoforming polypropylene. If the sheet temperature is too low, the sheet will be too rigid to form and if the sheet temperature exceeds the melting point, the sheet may not have enough rigidity to support its own weight. It sags excessively in the thermoforming oven. It

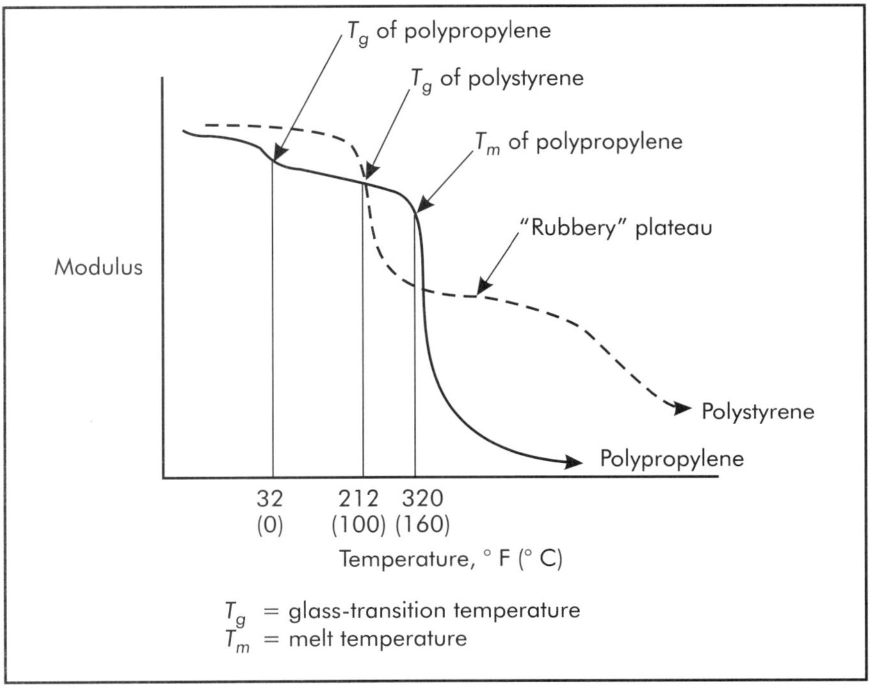


*Figure 2-6. Change of modulus with temperature. (Courtesy BP Polymers)*

should be noted that although the modulus change in polypropylene at the glass-transition temperature is relatively small, this temperature causes a large change in toughness.

For an amorphous material such as polystyrene, a major drop of modulus occurs at the glass-transition temperature. Due to entanglements between the molecules, a polystyrene sheet heated above its glass-transition temperature still has reasonable stiffness and behaves like a leathery or rubbery material. At this point it is soft enough to thermoform, but not so soft that it sags excessively. At higher temperatures, the molecules move rapidly and the polymer begins to behave like a viscous melt. The length of this rubbery plateau region depends on the molecular weight of the polymer. The plateau is broader at a higher molecular weight. Existence of this rubbery plateau means that an amorphous polymer generally is thermoformed over a broader temperature region when compared to a crystalline material.

## Heat-distortion Temperature

The heat-distortion temperature (HDT) is a measure of the temperature that a material can withstand while retaining a reasonable level of stiffness. This is also measured using the three-point bend test, as for flexural modulus. However, here the force is kept constant and the temperature is gradually increased until the beam deflects by 0.100 in. (2.54 mm) at its midpoint (see Figure 2-5). Typical HDT values are reported at two different values of flexible stress, 66 psi and 264 psi (455 kPa and 1,820 kPa).

From a material standpoint, the HDT value is affected by many of the same factors that influence the modulus. For amorphous polymers, the HDT value is generally close to the glass-transition temperature, and both values increase as the polymer backbone becomes more rigid. For semicrystalline polymers, the HDT lies between the glass-transition temperature of the amorphous phase and the melting point of the crystalline phase. As crystallinity increases, the HDT also increases. When a semicrystalline material is reinforced with a hard filler such as fiberglass or talc, the HDT can be increased significantly and may lie very close to the melting point.

In terms of processing conditions, increasing the orientation of a part will generally reduce its performance at elevated temperatures, since the part will tend to warp and distort as noted previously. The reported HDT values are obtained on injection mold specimens, and this data cannot be used in general to provide guidance on orientation effects.

## Toughness

For many applications, a thermoformed part must have sufficient toughness and ductility to withstand high-speed impact or large-scale bending deformation. For rigid food packaging, the container should not break when dropped. If the package is exposed to refrigerator or freezer temperatures, the impact properties may be even more critical, since many plastics become more brittle at low temperatures. For amorphous materials, the impact behavior ranges from very brittle for crystal polystyrene, to very tough and ductile for polycarbonate. The impact strength of polystyrene can be dramatically improved by blending a rubber in with the polymer. The rubber-modified version of polystyrene is known as high-impact polystyrene (HIPS) and typically contains 7–10% of a polybutadiene rubber dissolved in the styrene monomer before it is polymerized. This rubber lowers the stiffness of the polystyrene and causes it to become opaque.

Among crystalline polymers, toughness generally increases as the level of crystallinity decreases. This leads to a trade-off between the stiffness and impact strength of the material. Polypropylene (PP) is noteworthy in that it becomes very brittle at low temperatures (below 32° F or 0° C), due to the fact that the glass-transition temperature of the amorphous phase is very close to 32° F (0° C). The toughness below $T_g$ can be improved by using polypropylene impact copolymer (ICP), which contains small rubber particles consisting of an ethylene-propylene copolymer, dispersed throughout the matrix in the homopolymer phase. These ICPs are usually made by having two reactors in series with one another at the polypropylene production plant. PP homopolymer is made in the first reactor, and the ethylene-propylene rubber is made in the second. Rubber can also be blended with PP homopolymer in a separate compounding operation.

The dispersed rubber particles have a much lower $T_g$ than the matrix (often as low as –58° F or –50° C) and they act to absorb energy during an impact event. Increasing the level of rubber will increase the toughness of the resin, but will also reduce its stiffness. As is the case with HIPS, addition of rubber to PP causes the resin to become more opaque.

Another way to increase the toughness of polymers is by increasing their molecular weight (or lowering the melt flow rate). Longer molecules participate in more entanglements, thereby increasing the integrity of the entangled polymer network. In semicrystalline materials, the molecules that connect small crystals to one another are often the weak link in the structure, and the number of tie molecules increases with increasing molecular weight. From a processing standpoint, increasing the molecular weight of the polymer causes it to become more viscous in the molten state, and therefore more difficult to extrude into sheets or inject into parts molds. A trade-off therefore exists between processability and toughness with regard to the molecular weight of the material.

Another factor that can effect toughness is the orientation present in the test specimen. Some brittle polymers, such as crystal polystyrene, become much tougher when oriented. This feature is used in making thermoformed parts from an oriented polystyrene sheet (OPS). Here, the extruded PS sheet is first passed over a series of heated rollers, with the downstream rollers turning at higher speed. This causes the sheet to orient in the machine direction. The sheet then goes into a heated oven known as a tenter frame, where the edges of the sheet are gripped by a series of metal clips attached to moving chains. Rails that carry the chains diverge, causing the sheet to be stretched in a cross or transverse direction. The final sheet, which is biaxially oriented (oriented in two directions), is thermoformed, producing very tough, oriented parts.

## Other Important Physical Properties

In addition to mechanical properties, important physical and appearance properties can determine the choice of materials for a given application. Examples of physical and appearance properties include barrier capabilities, clarity, gloss, creep, stress and fatigue resistance, and long-term degradation.

### Barrier

For food packaging, the freshness and shelf life of a product often are determined by how effective packaging is as a moisture or oxygen barrier. These properties are measured by oxygen permeability and by the moisture vapor transmission rate (MVTR).

Some materials, such as polypropylene, are excellent moisture barriers (low MVTR), but are not very good oxygen ($O_2$) barriers. The $O_2$ permeability of a PP container can be improved considerably by coextruding a multilayer sheet containing a barrier layer of a low $O_2$ permeability material such as ethylene-vinyl alcohol (EVOH).

## Clarity

If high clarity or transparency is required, it is important that the sheet contain only a single homogeneous polymer phase, without any particles of a second phase that can scatter light. Amorphous polymers, such as polystyrene, are quite clear. However, when dispersed rubber particles are present such as in HIPS, the material is opaque. Crystalline polymers, such as PP, range from translucent to relatively clear depending on the size of the spherulites. As noted earlier, addition of a nucleating agent to the PP greatly reduces spherulite size, thereby improving clarity. If a low level of ethylene is copolymerized with propylene to make a random copolymer (RCP), crystallinity is reduced and clarity improves. However, the drawback to this approach is that the RCP also has a reduced modulus, heat-distortion temperature (HDT), and melting point relative to the homopolymer PP.

The effect of crystallinity on clarity and other end-use properties is seen dramatically in polyethylene terephthalate (PET). Since PET crystallizes more slowly than other semicrystalline materials, such as PP and HDPE, it is possible to produce an extruded PET sheet that is completely amorphous. If a cold mold is used, it is possible to maintain the amorphous state throughout the forming process to produce a clear amorphous polyethylene terephthalate (APET) part. If a hot mold is used and there is sufficient residence time in the mold, the polymer will crystallize (CPET), forming an opaque part. Due to the high melting point of crystalline PET, the CPET parts have excellent high-temperature stability and can be used for applications such as ovenable dinner trays. CPET parts also are microwavable.

## Gloss

High surface gloss is a requirement for many products, especially where the product is printed. Gloss is affected by both composition of the polymer and processing conditions. For a part surface to exhibit high gloss, the surface must be as smooth as possible. Any surface roughness leads to light scattering, thereby reducing surface gloss. Because of this scattering, many factors that influence clarity also influence gloss.

Amorphous materials can achieve a high gloss level. If dispersed rubber particles are present, such as in high-impact polystyrene (HIPS), surface gloss is reduced. HIPS generally contains smaller rubber particles and thus higher gloss levels are possible. Acrylonitrile butadiene styrene (ABS), which also contains dispersed rubber particles, gives very high gloss since the rubber particles are much smaller than the wavelength of light, and therefore scatter little light.

For semicrystalline polymers, the spherulite structure can lead to surface irregularities that scatter light and reduce gloss. Since nucleation reduces spherulite size, it also improves surface gloss. Processing changes also can be used to reduce spherulite size and improve gloss. If cold-polishing rolls are used to quench the molten sheet, the sheet surface will contain smaller spherulites and have improved gloss.

It is possible to produce a glossy part from a nonglossy material such as HIPS by coextruding a skin layer of crystal polystyrene. In a similar manner, a nucleated PP skin can be coextruded over a non-nucleated core layer to improve surface gloss.

## Long-term Properties

In certain applications, a thermoformed part may be exposed to a hostile or aggressive environment for extended periods of time, or forced to withstand high and/or varying levels of stress and deformation. It is important to know how the material will respond to this type of situation, and to predict its useful service life.

In the case of mechanical stress, especially at elevated temperatures, the creep properties of the polymer must be known. *Creep* refers to the slow change in dimensions that occurs as a function of time when the plastic is exposed to a constant force or stress.

For some plastics, exposure to certain environmental agents that normally have no adverse effect on a stress-free part can cause catastrophic failure when the agents contact a part under mechanical stress. This property, known as *environmental stress-crack resistance* (ESCR), is exhibited by polystyrene when the environment agent is a grease or oil, such as butter or margarine. In the case of high-density polyethylene, exposure to detergent solutions also can lead to stress-crack failure. This type of failure can even occur in the absence of external mechanical stress if the exposed part contains a high level of internal stress or orientation caused by the forming process. This is a critical concern for detergent bottles made from HDPE.

Another type of long-term property is *fatigue resistance*. Sudden catastrophic failure can occur to a part subjected to many loading cycles, such as a container hinge. Polypropylene is known to exhibit a living-hinge property, which makes it very resistant to this type of failure, provided the hinge is pre-bent soon after the part is formed.

Many polymers undergo long-term degradation of their properties when they are exposed to agents such as heat, light, radiation, oxygen, ozone, and certain chemicals. Unlike ESCR, this type of aging does not require a simultaneous stress on the part. Antioxidants and other stabilizers are typically added to many polymers to minimize oxidative and thermal degradation that can occur during melt processing and end-use, for example, spin baskets in washers and dryers. Since high levels of regrind from the ground-up skeleton sheet are often used to make thin-gage thermoformed parts, the drop in viscosity (rise in MFR) due to the degradation of polypropylene can be a major concern because it leads to increased sag.

It also is not uncommon to observe that, when certain plastics are left out in the sun for extended periods of time, they become brittle and eventually lose most of their mechanical strength. This degradation occurs because solar radiation causes polymer chains to split, thereby reducing molecular weight. Sunlight also can cause the chains to cross-link, thereby reducing ductility. Ultraviolet (UV) radiation generally is most damaging. In many cases, ultraviolet stabilizers can be incorporated in the plastic to minimize these degradation reactions. Adding certain pigments to the plastic, such as carbon black or $TiO_2$ (white), also can screen out UV rays.

In plastics used for medical applications, parts often need to be sterilized, and nuclear radiation is often used to accomplish this.

These nuclear rays (such as cobalt 60 radiation) usually cause the same type of degradation produced by UV radiation, and similar types of stabilizers have been used effectively to protect the plastic. Table 2-1 shows a comparison of thermoforming sheet properties.

## Flow Behavior of Polymers

The first step in producing a thermoformed part is the production of an extruded sheet. The extruder melts and mixes the polymer so that a uniform sheet can be extruded from the die. There are many factors that must be considered in terms of controlling and optimizing the flow of the polymer in this process.

The heat used to melt the polymer in the extruder is supplied by both electrical heaters that surround the barrel and friction caused by shearing the polymer between the barrel and screw. This mechanical energy often exceeds that supplied by the electrical heaters. The work required to pump polymer through the extruder is determined by its viscosity, which is a measure of the force required to make it flow. Virtually all fluids exhibit some resistance to flow that must be overcome by force or pressure. It is a common experience that thick fluid with a high viscosity, such as oil or molasses, is more difficult to pump through a tube when compared to a low-viscosity fluid like water.

One way to measure viscosity is to apply pressure to the fluid in a tube and measure the flow rate this produces. For simple fluids, such as water, flow rate out of the tube doubles if pressure is doubled. The viscosity of fluid is proportional to the ratio of pressure to the flow rate, and this ratio remains constant with changes in either pressure or flow rate. Liquids that obey this relationship are referred to as *Newtonian.* Temperature also affects viscosity, and most fluids become thinner and less viscous when the temperature rises.

For polymers, the situation is more complicated due to the entangled nature of the molecules. A standard way of measuring the viscosity of a polymer is the melt-flow rate (MFR) test (see Figure 2-7).

Table 2-1. Comparison of thermoforming sheet properties (Courtesy KAMA Corp.)

| Test | OPS | HIPS | HDPE | PP | APET/RPET | PETG | RPVC |
|---|---|---|---|---|---|---|---|
| Maximum tensile strength, psi (MPa) | 8,000–12,000 (55–83) | 4,000–6,000 (28–41) | 4,000–6,000 (28–41) | 4,000–6,000 (28–41) | 6,000–8,000 (41–55) | 6,000–8,000 (41–55) | 6,000–8,000 (41–55) |
| Maximum flexural modulus, psi (MPa) | 420,000–475,000 (2,896-3,275) | 325,000–375,000 (2,241–2,585) | 200,000–300,000 (1,379–2,068) | 200,000–300,000 (1,379–2,068) | 250,000–350,000 (1,724–2,413) | 200,000–300,000 (1,379–2,068) | 350,000–400,000 (2,413–2,758) |
| Maximum elongation, % | 3–50 | 30–50 | 200–400 | 100–400 | 100–300 | 200–400 | 30–200 |
| Maximum tear strength, lb/in. (N/m) | 1.1–1.5 (193–263) | 1.5–2.5 (263–437) | 2.0–3.0 (263–525) | 2.0–3.0 (263–525) | 1.5–2.5 (263–437) | 2.0–3.0 (263–525) | 1.5–2.5 (263–437) |
| Maximum folding endurance | F | G | G | E | G–E | E | Good |
| Minimum water absorption, % in 24 hr. | 0.04–0.1 | 0.04–0.1 | 0–0.01 | 0–0.005 | 0.5–1.0 | 0.3–0.8 | 0–0.005 |
| Minimum water vapor transmission | 7.0–10.0 | 7.0–10.0 | 0.25–0.35 | 0.65–0.75 | 2.0–4.0 | 3.0–5.0 | 4–6 |
| Gas transmission | | | | | | | |
| Carbon dioxide | 900 | 900 | 580 | 500–800 | — | — | 100–300 |
| Hydrogen | — | — | — | 1,700 | — | — | — |
| Nitrogen | — | — | 42 | 40–48 | — | — | — |
| Oxygen | 350 | 350 | 185 | 150–240 | 5–15 | 20–30 | 35–190 |

Table 2-1. (continued)

| Test | OPS | HIPS | HDPE | PP | APET/RPET | PETG | RPVC |
|---|---|---|---|---|---|---|---|
| Maximum chemical resistance | | | | | | | |
| Strong acids | G | G | E | E | G | G | G |
| Strong alkalies | E | E | E | E | G | G | G |
| Grease and oils | G | G | G–E | G | E | E | G |
| Organic solvents | E–P | E–P | G–E | G–E | E | E | G |
| Water | E | E | E | E | E | E | E |
| Maximum sunlight | F | F | F–G | F | G–E | G–E | G |
| Thermoforming conditions | | | | | | | |
| Upper limit, ° F (° C) | 330 (166) | 330 (166) | 360 (182) | 360 (182) | 320 (160) | 320 (160) | 300 (149) |
| Normal limit, ° F (° C) | 290 (143) | 295 (146) | 295 (146) | 310 (154) | 300 (149) | 300 (149) | 245 (118) |
| Lower limit, ° F (° C) | 270 (132) | 260 (127) | 260 (127) | 260 (127) | 250 (121) | 250 (121) | 200 (93) |
| Set temp., ° F (° C) | 170 (77) | 165 (74) | 170 (77) | 165 (74) | 155 (68) | 150 (66) | 150 (66) |
| Minimum shrinkage in the mold, in./in. (mm/mm) | 0.004–0.006 | 0.004–0.006 | 0.015–0.020 | 0.010–0.018 | 0.002–0.004 | 0.003–0.005 | 0.004–0.006 |
| Upper temperature limit (max.), ° F (° C) | 175 (79) | 165 (74) | 225 (107) | 240 (116) | 155 (68) | 150 (66) | 145 (63) |
| Lower temperature limit (min.), ° F (° C) | 0 (–18) | –20 (–29) | –150 (–101) | 0 (–18) | 0 (–18) | –20 (–29) | –20 (–29) |
| Maximum clarity | E | P | P | G | G–E | G–E | F–G |
| Maximum light transmission, % | 90–92 | 10–50 | 10–30 | 50–80 | 85–92 | 85–92 | 60–80 |

APET = amorphous polyethylene terephthalate
OPS = oriented polystyrene sheet
RPET = recycled polyethylene terephthalate
HDPE = high-density polyethylene
PETG = polyethylene terephthalate glycol
RPVC = rigid polyvinyl chloride
HIPS = high-impact polystyrene
PP = polypropylene
E = excellent F = fair G = good P = poor

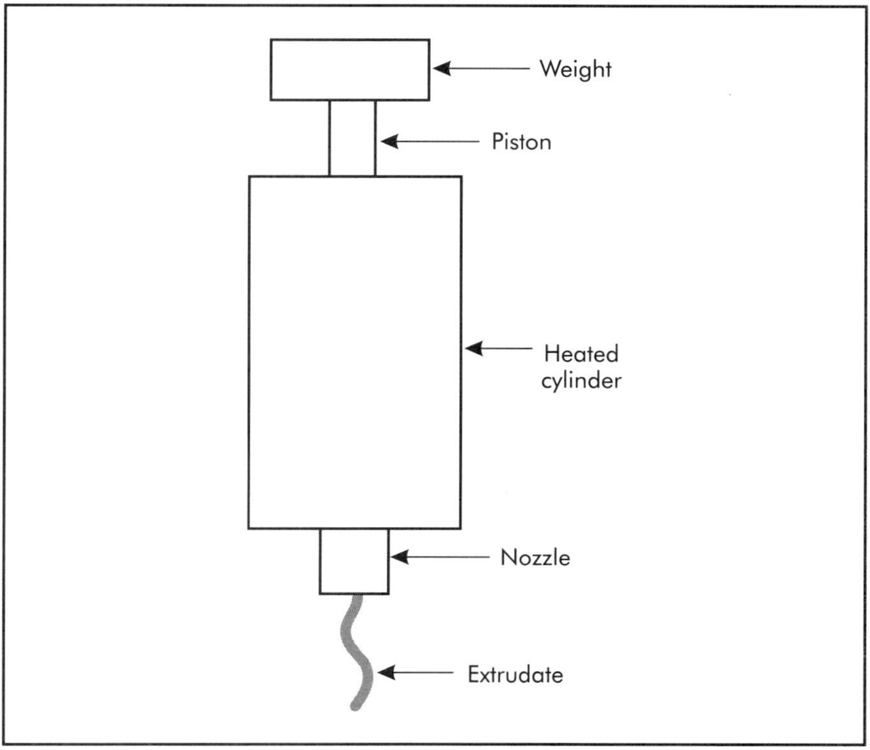


*Figure 2-7. Melt-flow rate (MFR) test (Bryce 1998).*

## Extrusion

In Figure 2-7, the polymer pellets are melted in a heated barrel and a weighted plunger is used to force melt out of a die at the bottom of a barrel. The melt-flow rate (MFR) represents the number of grams of polymer that extrudes from the die during a 10-minute time period using a standard weight (for example 4.761 lb or 2.16 kg as specified by ASTM D1238) on the plunger. The higher the viscosity of the polymer, the lower the MFR.

For most polymers, viscosity does not remain constant as the pressure or extrusion rate changes. For example, if the weight on the polymer is doubled, pressure doubles, but flow rate usually goes up by more than a factor of two. In other words, as extrusion rate increases, viscosity decreases. In this case, the ratio of pressure to flow rate is not constant. This type of flow behavior is called *non-Newtonian*. This also is referred to as shear-thinning behavior.

*Shear rate* is the term used for influencing viscosity. A fluid is sheared when it moves past a stationary surface, such as the wall of a die or the barrel of an extruder. Shear rate is related to how rapidly this movement occurs. Shear rate increases as the flow rate of fluid through the die increases, or if size of the die hole (or die gap in a flex-lip die) decreases. The shear-thinning behavior of polymers means that viscosity decreases as shear-rate increases. This shear-thinning behavior has important practical implications involving such things as pressure build-up in the die. If two resins have the same MFR (which is measured at low shear rate), but one material shear thins to a greater extent, the more shear-sensitive resin is less viscous at a high shear rate. During extrusion, this less viscous resin will exhibit a lower die pressure at a given extrusion output rate, or extrude at higher rates at a given pressure.

Figure 2-8 shows flow behavior of three different polypropylene resins, two of which have the same MFR but different shear sensitivities (B and C), and one that has a higher MFR (A). The more shear-sensitive material (B) has a lower viscosity at high shear rates. Shear rates in this higher range are seen in the extruder, and lower viscosity means the more shear-sensitive material extrudes more easily, allowing higher output rates and/or lower

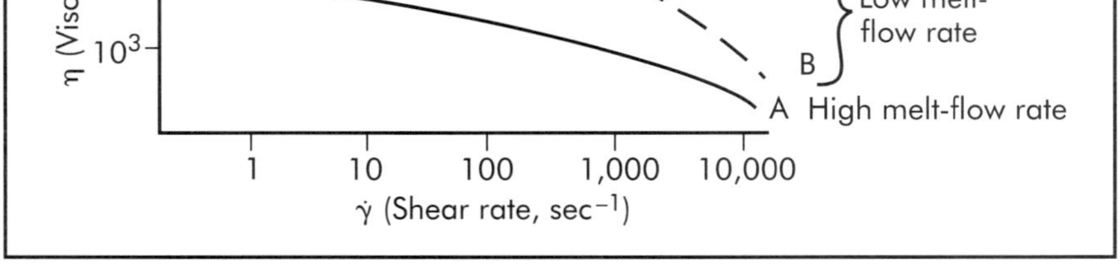


*Figure 2-8. Flow behavior of three different polypropylene resins. (Courtesy BP Polymers)*

pressures. At low shear rates, this material is more viscous; it has higher melt strength and exhibits less sag. The high-MFR resin (A) shows the most sag during thermoforming. The shear sensitivity of a polymer is often controlled by its molecular-weight distribution. The greater shear sensitivity of resin B is due to the fact that it has broader molecular-weight distribution when compared to resin C.

Shear thinning is due to the entangled nature of the polymer melting. As shear rate increases, long polymer strands line up in the direction of flow and disentangle from one another. These oriented, disentangled polymer chains offer less resistance to flow, causing the fluid to have a lower melt viscosity.

Increasing the melt temperature reduces the viscosity of polymers, but this temperature sensitivity differs for different polymer types. In general, polymers with stiffer backbone chains (higher glass-transition temperatures) show a greater drop in viscosity with temperature compared to more flexible materials. For example, a 50° F (28° C) change in melt temperature for a typical propropylene will cause melt viscosity at low shear rates to drop about 15%.

## Melt Elasticity, Melt Strength, and Orientation

The long-chain nature of polymers causes molten material to behave in some respects like an elastic solid when it is subject to external forces. This means that if a molten polymer is subjected to a sudden stretching force, it tends to snap back when this force is released. If the force is applied slowly over a longer period of time, the polymer flows like a viscous liquid. This type of behavior is easily seen with Silly Putty™, which is a silicone polymer. When a ball of Silly Putty is dropped onto a hard surface, it bounces like a rubber ball (elastic). When this same ball of Silly Putty is placed on a surface for a long period of time, it gradually flows into a puddle under the influence of gravity (viscous). For this reason, polymers often are referred to as being *visco-elastic*, since they have both viscous and elastic properties. This has important consequences in both sheet extrusion and thermoforming operations, and can also affect the final properties of the thermoformed part.

When melt is forced through the die, the molecular chains are forced to disentangle and align in the flow direction. After the melt exits the die, the chains try to go back to a more random, entangled arrangement. In some ways, this is like stretching a rubber band and then allowing it to spring back. The springback of the melt causes the bead of molten material at the die lip exit to thicken; this is referred to as *die swell*. It is easy to measure the extent of die swell by running a melt-flow-rate test on the polymer and quenching the molten strand in water as it exits the die so that the weight of the strand does not cause it to draw down. The thickness of the solid strand will be significantly larger than the die opening. The degree of die swell depends on the processing conditions used (shear rate and temperature) and the molecular weight distribution of the polymer. Higher shear rates cause die swell to increase. Broader molecular weight distributions and higher molecular weights also cause an increase in die swell. Die swell decreases with increasing temperature.

If the shear rate or molecular weight gets very high, or melt temperature gets very low, the extrudate not only swells, but also may distort or even fracture. This is referred to as *melt fracture*, and the extrudate may exhibit either regular or irregular thickness changes (see Figure 2-9). When this occurs during sheet extrusion, the surface of the sheet may take on an undesirable surface texture like that of an orange peel. To correct this problem, either shear rate or melt viscosity of the polymer must be lowered. Reducing the extruder rpm (output) or increasing die gap can lower shear rate. Raising melt temperature or going to a higher melt-flow-rate material can reduce the melt viscosity of the polymer.

When the extruded sheet is reheated in the oven before being formed, its temperature must be raised to the point where it is soft enough for forming to occur. For amorphous materials, this requires a temperature above the glass-transition temperature, while for crystalline polymers, the sheet temperature must often equal or exceed the melting point of the polymer. For amorphous materials, the softening process occurs gradually as the glass-transition temperature is exceeded. The sheet goes through a broad transition where it has an almost rubber-like or leather-like texture. This is ideal for thermoforming, since the sheet is soft enough

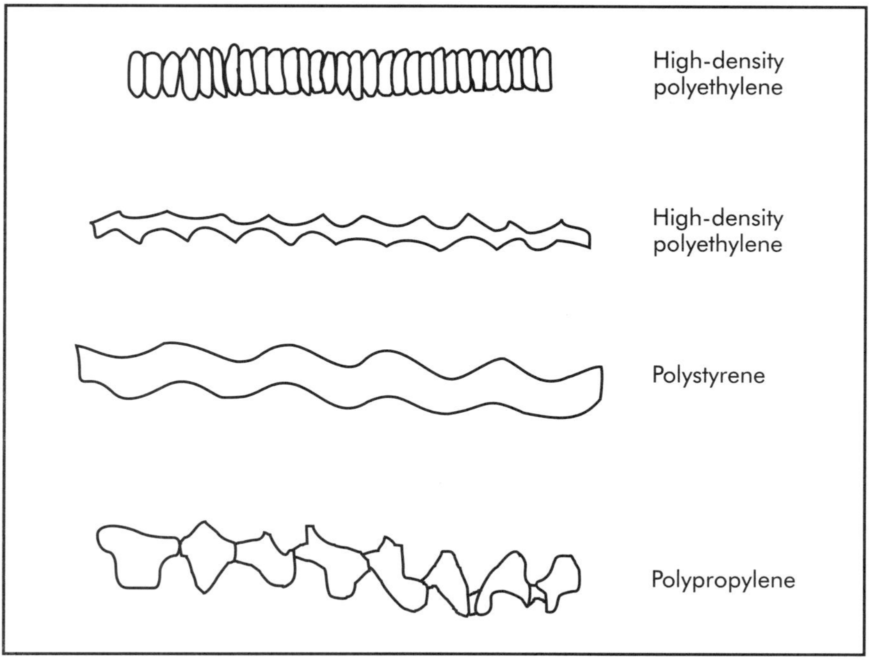


*Figure 2-9. Extrudate may exhibit either regular or irregular thickness changes due to melt fracture. (Courtesy BP Polymers)*

to be forced into the mold using reasonable pressures, but is not yet a fluid melt, which would sag in the oven under its own weight.

For crystalline polymers, such as polypropylene or polyethylene, transition from a solid sheet to a fluid melt is much more abrupt and can occur within a span of just a few degrees when the melting point is reached. Sheet sag is therefore a major concern for crystalline materials and melt strength is very important. *Melt strength* basically refers to the degree to which the molten sheet resists thinning down under the influence of gravity when it is in an unsupported state. The melt strength increases as the molecular weight of the polymer increases. For polypropylene, very low MFR or even fractional MFR materials have been used when high melt strength is required. One drawback of these materials is a reduction in extrusion output occurring at a very high molecular weight and an increasing incidence of melt fracture.

Melt strength also can be increased by a change in molecular architecture through introduction of long chain branches. This type of branching often occurs in polyethylene during extrusion, so use of high levels of PE regrinds generally results in an increase in melt strength. For polypropylene, on the other hand, extrusion generally leads to a reduction in molecular weight (rise in MFR) without formation of branches, and this causes reduced melt strength when high levels of regrind are used.

When the molten sheet exits the die and contacts the polishing roll stack, takeoff speed is adjusted to the desired final sheet thickness. Increasing the takeoff speed of the rolls causes the sheet to become thinner, and also increases tension on the molten sheet. This tension causes the molecules to orient in the machine direction. When the sheet contacts the polishing roll, it begins to cool and solidify. Some of this orientation is frozen in since molecules are no longer free to move around.

When the sheet is heated back up in the oven prior to thermoforming, molecules regain their freedom of motion, and this causes the sheet to shrink and distort. High or nonuniform levels of machine direction orientation are undesirable, since they lead to large dimensional changes in the sheet. One exception to this statement is the situation of oriented polystyrene sheet (OPS), where biaxial orientation (in both the machine and transverse directions) is deliberately induced before the extruded sheet passes into the reheat oven. This orientation leads to final thermoformed parts that are less brittle than when unoriented polystyrene (PS) sheet is used.

One final place where orientation is induced is the forming process itself. If the sheet is in the melt state during forming, some relation of this orientation may occur before the part solidifies. For some materials, such as polypropylene (PP), which has poor melt strength, the sheet may be formed in solid state through the use of very high pressures. This results in high levels of stress and orientation in the final part, and limits high-temperature use of the part (for example, as a container for microwave heating of foods) due to distortion that occurs during reheat. One interesting product benefit associated with solid-state forming of PP is the high level of clarity achieved in the final parts. This is due to the fact that the drawing, which occurs when the solid sheet is formed, destroys the spherulitic structure responsible for light scattering.

## Choosing a Resin

Several choices of plastic usually are available once the physical characteristics of the resin needed for a particular thermoformed component are determined. An economic comparison can be made among them, as shown in Table 2-2.

The modulus of elasticity governs the rigidity of similar geometric shapes as illustrated in data on the line labeled "Equal performance thickness" in Table 2-2. When this factor is applied in conjunction with yield, an economic determination can be made. The final cost per 1,000 trays, assuming 20% return on web scrap, indicates the cost differences of using various resins.

From Table 2-2, the final cost of the clear sheet per 1,000 trays is as follows: PVC = \$36.03; APET = \$35.28; RPET = \$28.26; and OPS = \$24.35. The obvious choice should be OPS if cost is the only consideration. However, OPS is more economically processed on a special type of thermoformer rather than the more available radiant heat equipment found in most plants. RPET is cost-attractive, but being a recycled material it may not always be readily available. Polyvinyl chloride (PVC) is a very popular clear resin, but it can cause an environmental problem when discarded at solid-waste disposal sites. APET is often substituted for PVC, but some thermoforming operators claim to have problems trimming it with knife-like dies. The end user may have experienced some of these as well as other disadvantages that guide decision making.

The least expensive opaque sheet is polystyrene foam (used for egg cartons, etc.), which has a yield 5–15 times greater than its other competitors, but has very poor physical properties. The foam rolls are bulky and skeleton scrap shot must be recycled internally as the scrap has no commercial value and cannot be disposed of at a solid waste dump.

The thermoformability of a resin (ease of processing) is not solely based on its inherent physical properties, but also on external factors. Major influences are the design features and efficiency of

Table 2-2. 1999 Economic comparison of resins (Courtesy KAMA Corp.)

| **Polymer Description** | **PVC** | **APET** | **RPET** | **HIPS** | **HDPE** | **PP** | **OPS** | **PETG** |
|---|---|---|---|---|---|---|---|---|
| Extruded sheet, \$/lb (\$/kg) | 0.77 (0.35) | 0.83 (0.38) | 0.67 (0.30) | 0.70 (0.32) | 0.75 (0.34) | 0.85 (0.39) | 0.76 (0.34) | 1.30 (0.59) |
| Specific gravity | 1.37 | 1.33 | 1.33 | 1.05 | 0.965 | 0.9 | 1.05 | 1.28 |
| Yield, in.$^2$/lb (cm$^2$/kg) at 10 mil (0.0254 mm) | 2,020 (28,280) | 2,081 (29,134) | 2,081 (29,134) | 2,636 (36,904) | 2,868 (40,152) | 3,076 (43,064) | 2,636 (36,904) | 2,163 (30,282) |
| Equal gage yield index, % | 100.0 | 103.0 | 103.0 | 130.5 | 142.0 | 152.2 | 130.5 | 107.0 |
| Cost/1,000 in.$^2$ (6,450 cm$^2$) | 0.3811 | 0.3988 | 0.3219 | 0.2655 | 0.2615 | 0.2764 | 0.2883 | 0.6012 |
| **Equal Gage Price Index, %** | | | | | | | | |
| PVC | **100** | 105 | 84 | 70 | 67 | 73 | 76 | 158 |
| APET | 96 | **100** | 81 | 67 | 66 | 69 | 72 | 151 |
| RPET | 118 | 124 | **100** | 82 | 81 | 86 | 90 | 187 |
| HIPS | 144 | 150 | 121 | **100** | 98 | 104 | 109 | 226 |
| HDPE | 146 | 153 | 123 | 102 | **100** | 106 | 110 | 230 |
| PP | 138 | 144 | 116 | 96 | 95 | **100** | 104 | 218 |
| OPS | 132 | 138 | 112 | 92 | 91 | 96 | **100** | 209 |
| PETG | 63 | 66 | 54 | 44 | 43 | 46 | 48 | **100** |
| **Equal Rigidity Performance** | | | | | | | | |
| Equal performance thickness, mils (μm) | 10 (254) | 9.5 (241) | 9.5 (241) | 10.6 (269) | 14.0 (356) | 13.3 (338) | 9.2 (234) | 10 (254) |
| Area per tray, in.$^2$ (cm$^2$) | 80 (516) | 80 (516) | 80 (516) | 80 (516) | 80 (516) | 80 (516) | 80 (516) | 80 (516) |
| Weight—lb (kg) per 1,000 trays | 39.60 (17.9) | 36.52 (16.6) | 36.52 (16.6) | 32.17 (14.6) | 39.05 (17.7) | 34.60 (15.7) | 27.92 (12.7) | 36.99 (16.8) |
| Sheet cost per 1,000 trays | 30.49 | 30.31 | 24.47 | 22.52 | 29.28 | 29.41 | 21 .22 | 48.09 |
| Sheet cost per 1,000 trays plus 20% web by weight | 36.59 | 36.37 | 29.36 | 27.02 | 35.14 | 35.29 | 25.46 | 57.71 |

Table 2-2. (continued)

| Polymer Description | PVC | APET | RPET | HIPS | HDPE | PP | OPS | PETG |
|---|---|---|---|---|---|---|---|---|
| **Equal Rigidity Performance (cont.)** | | | | | | | | |
| Web scrap value, $/lb ($/kg) | 0.07 (0.03) | 0.15 (0.07) | 0.15 (0.07) | 0.20 (0.09) | 0.15 (0.07) | 0.10 (0.05) | 0.20 (0.09) | 0.15 (0.07) |
| Final cost per 1,000 trays assuming Return on web scrap | 36.03 | 35.28 | 28.26 | 25.73 | 33.97 | 34.60 | 24.35 | 56.60 |
| **Equal Performance Price Index, %** | | | | | | | | |
| PVC | **100** | 98 | 78 | 71 | 94 | 96 | 68 | 157 |
| APET | 102 | **100** | 80 | 73 | 96 | 98 | 69 | 160 |
| RPET | 127 | 125 | **100** | 91 | 120 | 122 | 86 | 200 |
| HIPS | 140 | 137 | 110 | **100** | 132 | 134 | 95 | 220 |
| HDPE | 106 | 104 | 83 | 76 | **100** | 102 | 72 | 167 |
| PP | 104 | 102 | 82 | 74 | 98 | **100** | 70 | 164 |
| OPS | 148 | 145 | 116 | 106 | 140 | 142 | **100** | 232 |
| PETG | 64 | 62 | 50 | 45 | 60 | 61 | 43 | **100** |

Prices are for relative comparison only and are subject to change without notice. Thermoforming application rigidity index per Maurice Palmer Associates, Ltd.

APET = amorphous polyethylene terephthalate
OPS = oriented polystyrene sheet
PVC = polyvinyl chloride
HDPE = high-density polyethylene
PETG = polyethylene terephthalate glycol
RPET = recycled polyethylene terephthalate
HIPS = high-impact polystyrene
PP = polypropylene

thermoforming machines and the tooling used for production, plus the skill of operating technicians. Some resins require specialized tooling and equipment to achieve the highest economic efficiency. However, many of these plastics also can be processed on general-use machines at the cost of lowered productivity and quality.

Many older thermoforming machines can produce credible components at a lower production rate when processing high-impact polystyrene (HIPS) and PVC resins. These forgiving plastics are able to take a great deal of abuse from inexperienced machine operators and still thermoform saleable products. The same parts when run on modern equipment under the guidance of an experienced technician may show an increase in high-quality output by a factor of 100–500%.

Specialized thermoforming machines often are recommended for processing oriented polystyrene sheet (OPS) and foam polystyrene sheets. OPS sheet, which contains troublesome locked-in internal stresses that are released during oven heating, can make thermoforming difficult. This material is best processed on contact-heat pressure formers. Foam polystyrene has a very low density and therefore can cool much more rapidly in the mold than solid sheet styrene. It can be cycled at much faster rates than standard thermoforming machines can cycle. Machines specifically designed for high-speed thermoforming are built with low-weight platens. Careful control of acceleration and deceleration of the molds is exercised to process foam polystyrene.

Polypropylene requires machinery with very careful heating and thermal control of the sheet because of the narrow temperature range under which it can be thermoformed. As with all olefin resins, final part shrinkage may take up to 24 hours, which makes die-design shrinkage difficult to predict accurately. Builders of trim-in-place thermoformers claim an advantage in producing uniformly trimmed polypropylene parts because forming and trimming take place simultaneously.

Trimming of various plastics ranges from relatively easy (such as HIPS) to those plastics that may require special handling (olefins and the PET family). The choice of trimming, either using inexpensive knife-like dies or punch and dies, can strongly influence selection of resins because of the great difference in die costs. Knife-like die life can be prematurely shortened by abrasive or brittle plastics, which allow partial cutting and then crack away suddenly, causing the knife edge to pound into the striker plate.

The thermoforming technician, armed with knowledge of plastic materials and machinery peculiarities, often can rise above adversity and produce superior parts even with inferior equipment. The thermoforming industry should educate and retain these unusual and dedicated individuals because high-tech machinery and tooling alone cannot provide an answer for every forming problem.

## References

Bryce, Douglas, ed. 1998. *Plastic Injection Molding...Mold Design and Construction Fundamentals*. Dearborn, MI: Society of Manufacturing Engineers.

Mitchell, Philip, ed. 1998. *Tool and Manufacturing Engineers Handbook*, Fourth Edition, *Volume 8: Plastic Part Manufacturing*. Dearborn, MI: Society of Manufacturing Engineers.

# 3

# Design of Thermoformed Products

Thermoformed parts are designed primarily by the customer, package or product designer, and occasionally by the tool designer. Component design can match the needs of the customer and thermoformer if configured as a functional product adaptable for thermoforming. Plastic product customers and designers must be aware of the limitations of forming techniques to avoid production problems. There are few part-design guidelines, but each design may require a tradeoff of features that should be agreed to by all parties.

## Part Design Guidelines

Formed-part geometry helps determine the uniformity of wall thickness, corner strengths, ability to stack, and structural stiffness. It also can be the source of thermoforming flaws—such as holes, webbing, lack of clarity, or surface imperfections. If a product design were to call for a cubic shape (like dice) with sharp corners, without corner radii and zero draft, the result is predictable (see Figure 3-1). This part could contain holes, very thin walls, webbing, and severe sidewall thinning, because hot plastic material builds up on square corners instead of flowing uniformly over the mold. The edge at which sidewalls intersect near the mold top could induce webbing caused by the excessive heated plastic accumulating along the edge with insufficient mold surface available to absorb the excess material. Webbing is unsightly and is a

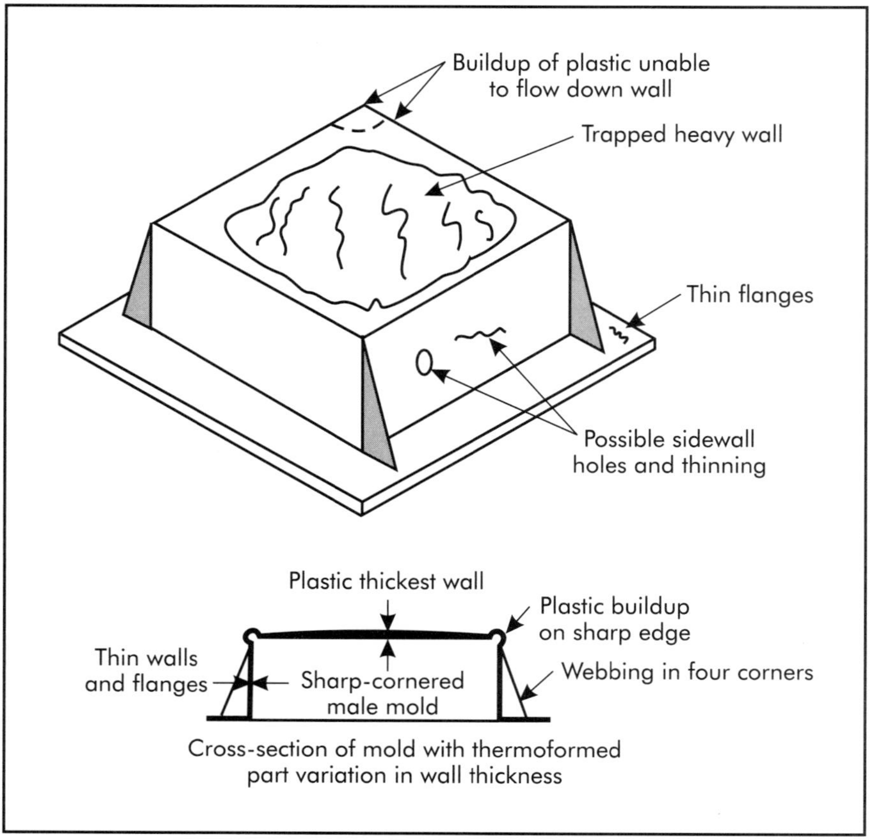


*Figure 3-1. Sharp-cornered rectangular part with zero draft produces defective thermoformed parts.*

significant weak point in part structure that prevents stacking the completed parts (see Figure 3-2).

## Modification

Modifications made to the cubical thermoformed part shown in Figure 3-3 can make it practical to thermoform. However, it still may not fully satisfy a customer requirement to tightly contain the packaged insert. Radii in all corners and the draft angle preclude a secure fit to the insert but are necessary for the thermo-

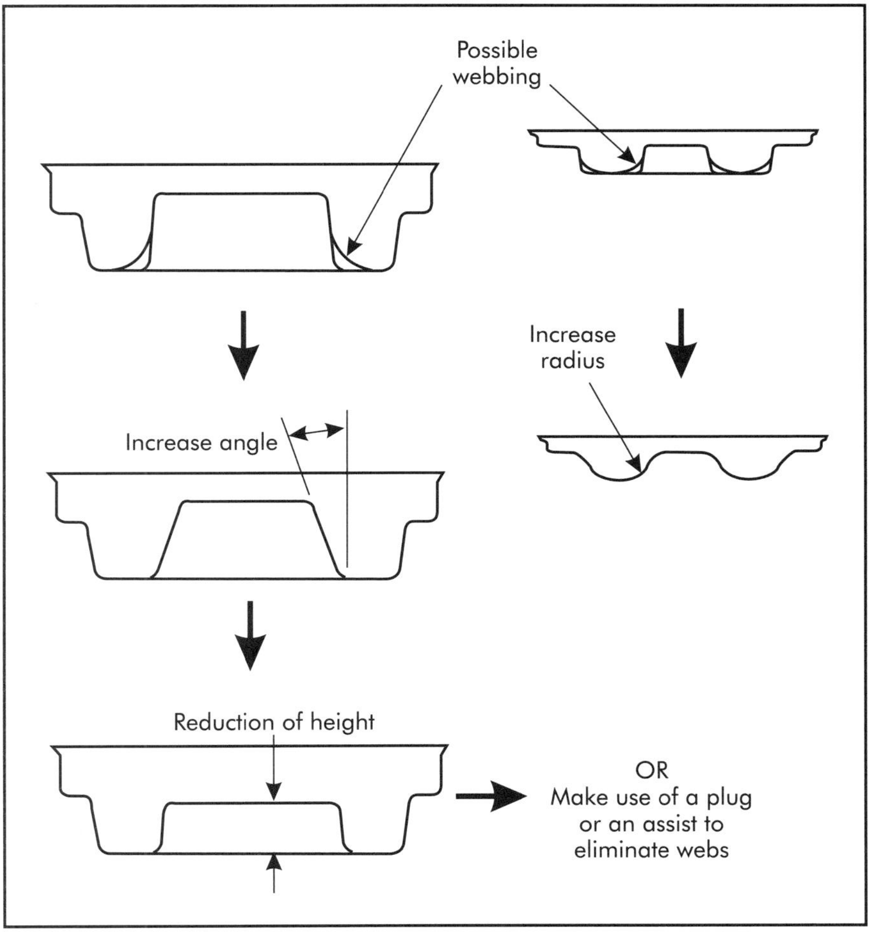


*Figure 3-2. Methods to reduce webbing. (Courtesy GN Plastics Co., Ltd.)*

forming process. A possible solution is further modification of the package design by introducing interior ribs and depressing the top surface panel as shown in Figure 3-4.

## Determining Proper Design Features

There are answers to questions like how large a radius the plastic should have or what the minimum angle of draft should be. The

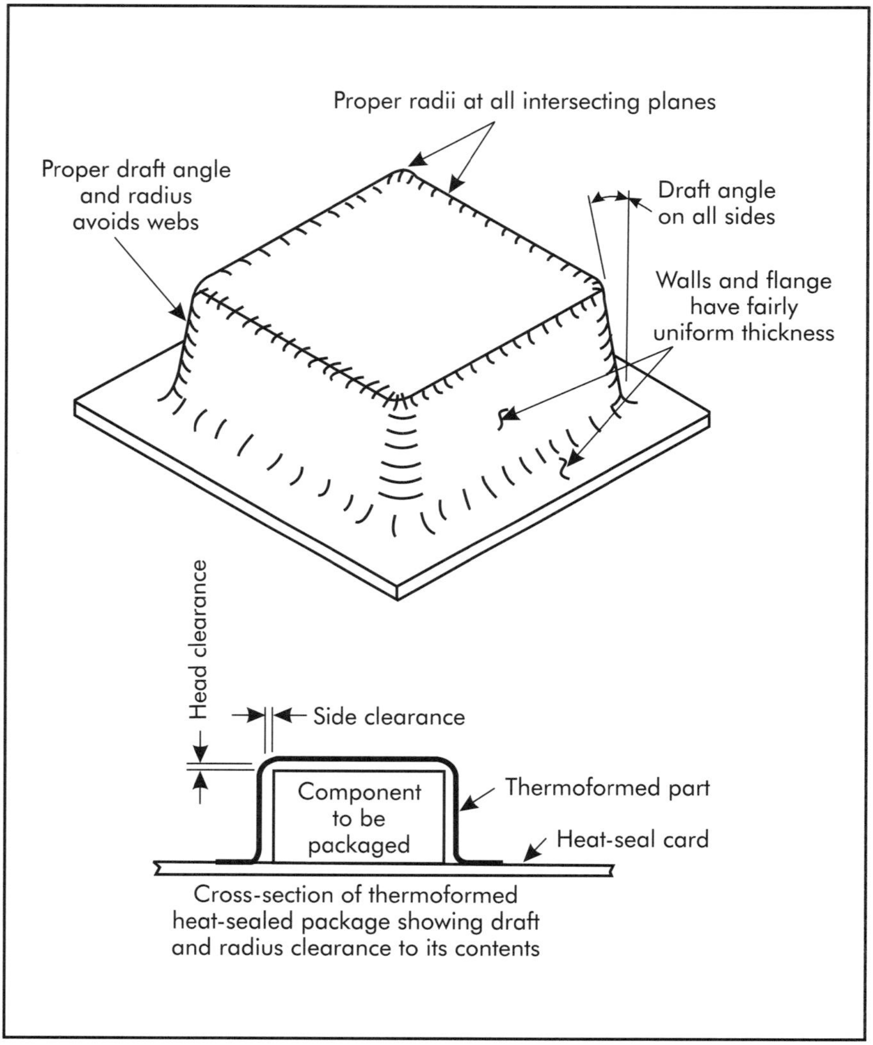


*Figure 3-3. A modified, thermoformed part with sufficient radius and draft angle may fit loosely to the packaged components.*

unsophisticated customer may specify dice-like formed parts. Often, the salesman who takes the order may be reluctant to ask for design revisions needed to assist in thermoforming. The tool designer should advise on these fundamental product-design deficiencies and advocate part revisions before starting work on a mold.

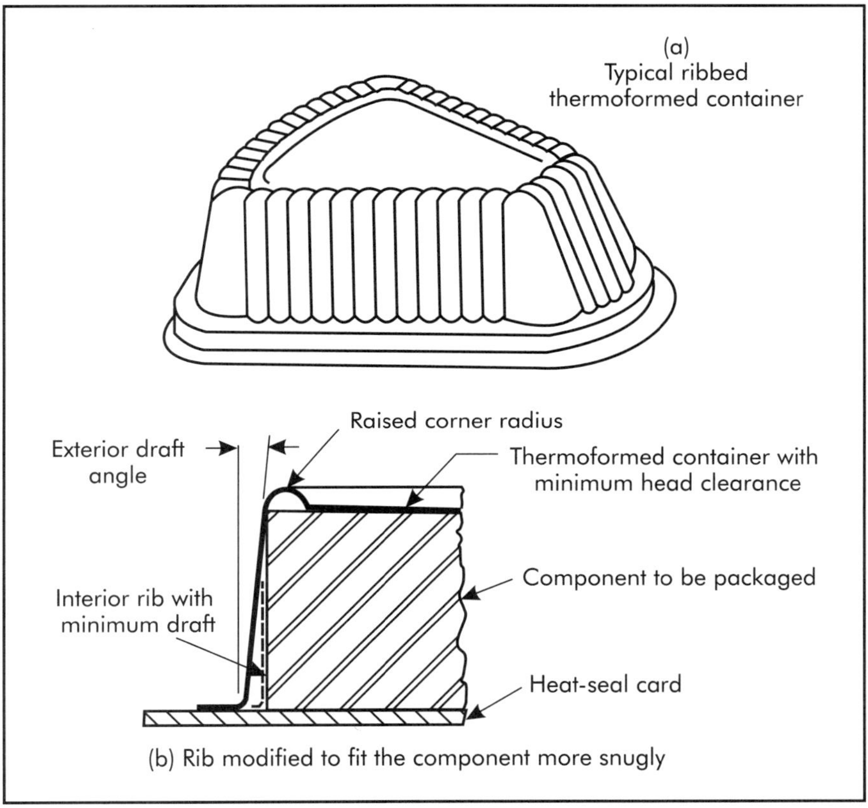


*Figure 3-4. Further modification of the package design by introducing interior ribs and depressing the top surface panel provide a snug fit to the product. (Courtesy GN Co., Ltd.)*

## *Draft Angle and Radii*

Providing sufficient draft angle controls some basic thermoforming parameters. Stripping parts from a male mold built with minimal draft can be difficult because hot plastic cools then shrinks and grips the cavity. Most parts with proper draft are easily removed from the mold by directing low air pressure between mold and plastic shot. At the other extreme, a large draft angle would self-strip parts without any assistance from air ejection.

Radii at all intersecting planes have a pleasing aesthetic purpose as well as being a functional necessity. The radius when presented to heated plastic allows it to flow smoothly (essential for uniform wall thickness) and reduces fracture due to stress concentration at the intersection of the planes. A minimum radius of 0.19 in. (4.8 mm) is a good starting point, and increasing the radius is preferable to reducing it.

### *Uniform Wall Thickness*

Consistent wall thickness reduces the weight of plastic required for a component. It is one of the most important objectives to be achieved when thermoforming. Many elements influence wall thickness, but after excluding most of them, it can be demonstrated that the greater the draft angle, the more uniform the wall.

### *Stack Height Variables*

The draft angle, component wall thickness, and any undercut projections govern stack height, which is the clearance between the flanges of two stacked parts (see Figure 3-5). The greater the draft angle, the closer together two parts can be stacked, resulting in a lower shipping cost. Increasing wall thickness for a given draft angle requires more separation between stacked parts. Experience indicates an initial design using a 5° draft angle should be adequate for most parts, although lesser angles are in common use. A smaller draft angle always requires tradeoffs with other thermoforming parameters.

### *Rib Stiffeners*

Rib stiffeners or texturing transforms a floppy flat-plane surface into a rigid structure able to take greater side and compression loading. A thinner sidewall can be specified if the product design permits—a layered surface or ribbed structure can provide the same rigidity as a thicker-wall, flat-plane surface. However, clear plastics, such as polyvinyl chloride (PVC) or oriented polystyrene sheet (OPS), sometimes are specified to provide maximum visibility into a package, and ribbing or layering may be an obstacle to clarity.

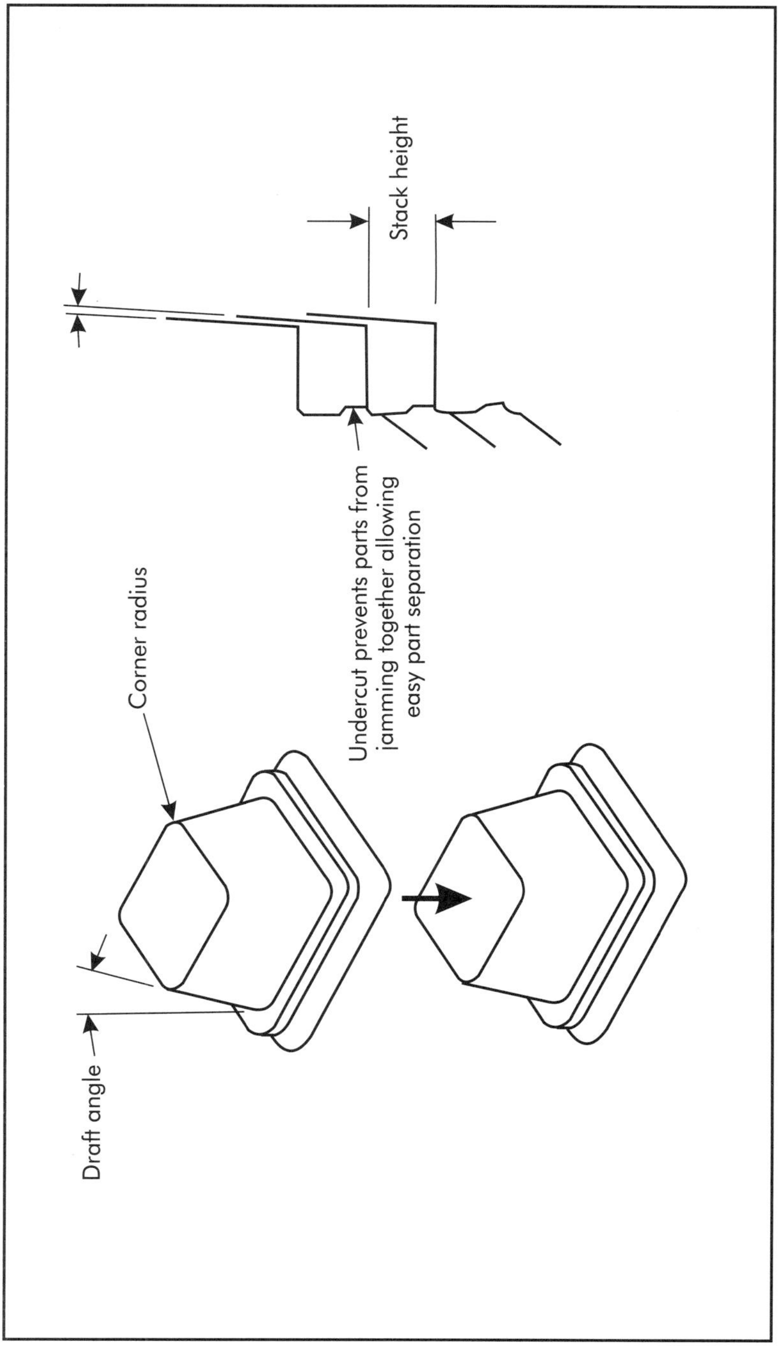


*Figure 3-5. Undercuts provide clearance between the flanges of two stacked parts. (Courtesy GN Plastics Co., Ltd.)*

Where food is eaten directly from a container with a spoon, package users prefer contoured, continuous surfaces without ribbing inside the package. Ribs are recommended if the goal is to secure a product firmly in place and add more protection against rough handling.

## Mold Texture

Mold surface texture can range from shiny smooth to leather-like, with many variations in between. It is essential to rapidly evacuate any entrapped air between the formed sheet and the cavity to prevent the sheet from cooling below its forming temperature. The rougher the mold surface texture, the more rapidly entrapped air can be vented to the atmosphere. Plastic resins such as polyethylene and polypropylene require an uneven mold surface to allow entrapped air to vent through the hills and valleys inherent in rough texture. Highly polished molds can require more vacuum holes or slot vents because entrapped air cannot travel very far along a smooth surface. Molds can be shot or sandblasted to help provide air passageways with the degree of roughness required for the type of plastic to be thermoformed.

Vent holes and slots may be inadvertently reproduced on the plastic surface of finished parts and may be found objectionable. A sandblasted cavity may exhibit a hazy surface on transparent plastic, which could result in loss of clarity. The diameter and number of vent holes, and width for evacuation slots, must be carefully matched to the product design, type of plastic being formed, and volume of air being vented in a specific time period. Experiments should be conducted on sample molds to produce a surface finish acceptable to the customer.

## Undercuts

Undercuts, as shown in Figure 3-5, are reverse-draft segments designed to retain a loose component temporarily, such as a lid on a cup. An undercut can provide a means of stacking thermoformed units so they will have a slight clearance between their sidewalls, preventing parts from jamming and sticking when they are dispensed. A plastic coffee cup lid is a very sophisticated example. Its outer and inner undercuts serve to snap into place on the cup. They also permit lids to be stacked quite high and yet be easily separated when dispensed.

The depth of undercut to design into the mold requires thought and testing. The finished result must be not too tight or too loose, but just right, which the designer will find is often a very subjective opinion. Undercuts can be designed so deep that they could require the mold to articulate to allow stripping the cavity—or the undercuts can be shallow enough to allow the part to contract away from a female mold wall and self-eject.

Teflon® coating on the mold makes stripping parts from a cavity easier by creating a slippery surface. Another technique is to use the front-and-rear sheet clamps on the mold base to eject shots. Undercut cavities that are difficult to strip may require mechanical knockout devices built into the mold system. A good starting point for an undercut segment would be 0.030-in. (0.76-mm) maximum depth below the cavity surface. If practical, the segment should be discontinuous to allow for easy removal.

## Inserts

The tool designer should study the loose insert that is to be snapped into an undercut part to achieve just the right fit. Cardboard inserts expand and shrink due to humidity, and the multiple-cavity steel rule used to trim out cardboard inserts has a tolerance of $\pm 0.010$ in. ($\pm 0.25$ mm). Thermoformed inserts formed against a mold side not facing the mating side of the corresponding part could have varying wall thickness, causing a dissimilar fit from part to part. These differences should be analyzed and errors reduced or otherwise compensated for before continuing design work.

Thermoformed sample parts, which are formed in the same manner as future production parts, should receive written approval from the customer to avoid costly misunderstandings caused by loose or tight fits. Because of the many variables to consider, conservative mold design dictates leaving extra metal to be machined away if the initial fit proves too tight.

## Trimming

Trimming out thermoformed parts from the shot is nearly always necessary, whether the formed shot contains single or multiple cavities. The tool designer must make thermoformed product decisions based on the trimming process to be employed and tolerances specified. Cut-in-place tooling offers the ability to hold the tightest tolerances of trim edge relative to thermoformed part geometry. For many reasons, this type of thermoforming process may not be available or suitable, so other trim methods should be considered.

On-line trimming using a knife-like die on a thermoformer with the die press situated on the machine frame requires the most generous trim-edge-to-formed-part tolerances.

The following variables in the use of knife-like dies should be considered when studying customer part specifications (see Table 3-1). The registration error from part-to-part could be as much as 0.030 in. (0.76 mm) for general industrial, knife-like, on-line die trimming. Careful control of variables can reduce the difference to 0.015 in. (0.38 mm). Tighter tolerances can be achieved by test running some production-trimmed sheets and building a new die to correct any inequalities observed in the trimmed parts.

**Table 3-1. Variables to consider when using knife-like dies**

| Specification | Variable |
|---|---|
| Die building tolerance | ±0.015 in. (±0.381 mm) |
| Machine consistency of index: old style equipment | ±0.030 in. (±0.762 mm) |
| Servo drive index equipment: modern machines | ±0.005 in. (±0.13 mm) |
| Ability to predict mold shrinkage of hot parts indexed into trim station | ±0.002 in./in. (±0.002 mm/mm) |
| Clearance to prevent knife from touching the plastic part and causing scuffing | 0.015 in. (±0.381 mm) |

Off-line trimming occurs when the trim press is detached from the thermoformer and a punch and die are employed to trim the plastic web, which has cooled and shrunk to final size. This process cuts only one or two rows of parts at a time from the shot, making it easier to align these rows to a die. Tolerances of the cut edge relative to the formed cavity can be controlled more accurately on off-line trim presses.

The trimming method used for a thermoformed component is a factor that must be decided by the part designer, end-product user, and thermoforming production management. The decision on the trim process to be employed can not be based solely on tolerance requirements. It should include decisions on die budgets, tool delivery, scheduling, and quantities to be produced. Part-trim tolerances are often loosened to suit a lower die cost and a more rapid delivery schedule when all facts are fully analyzed.

## Embossing and Texturing

Depending on the required result, several techniques are options for embossing and texturing plastic parts. Plastic sheet can be embossed during the extrusion process to simulate leather and wood grain. Other decorative finishes are available and the mold itself need not be textured. This technique is usually specified for parts that will not severely distort the decorative finish by elongating the pattern during thermoforming.

Molds can be chemically etched to provide a textured surface to a plastic thermoformed part formed from an unadorned plastic sheet. Photoengraved logos and lettering may be mounted as metal inserts in or on mold cavities. Mechanical engraving of the mold cavity produces deeper impressions than chemical etching, but at higher dollar cost.

Aluminum cast molds can be supplied with various decorative surfaces of fairly deep and complex designs. To do so, a pattern maker constructs a pattern with the necessary detail and mold shrinkage. Aluminum cavities are then cast in fine sand at foundries that specialize in thermoforming molds. This procedure can produce high-quality cavities at very low cost.

Either pre- or post-decorating can enhance or print information on thermoformed parts. Rolls of blank plastic can be prelaminated to color-printed-cover film, or the bare sheet can be decorated by laminating a film directly at the thermoforming machine. High-quality graphics directly printed on formed containers while using multi-color printing presses provide identification and advertising logos. This process takes place after the containers are trimmed. Containers are automatically fed to these machines, printed at high speed, and stacked for shipment.

## Thin-wall Formed Parts

Roll-fed, thin-wall thermoformed products are quite varied and are used by many industries in a large variety of forms, such as medical packaging, food containers, blisters, tray packaging, foam packaging, and continuous formed parts.

### Medical Packaging

Medical packaging used to protect sterile products is thermoformed and handled with special care in clean rooms built into the thermoforming factory. These clean rooms carefully filter incoming air and employees are uniformed and gloved with loose hair covered. The plastic materials are of the highest purity and quality. Tooling and machinery for medical components must be thoroughly inspected and cleaned so no loose plastic particulates or lubricants migrate into packaging. The greatest source of contamination often comes at the trimming operation, where dull or worn dies can cause fine angel hair or slivers of plastic to adhere to static-electrically charged packaging.

### Food Containers

Food containers must be kept clean and free of foreign matter. The volume of thermoformed lids, cups, plates, and food receptacles required is so large that these thermoforming plants use the latest automated equipment for production. It is not unusual to see production rates of up to 2,000-cups-per-minute per machine. Most cups manufactured in the United States are further processed by equipment that rolls the lips of the cups under their flanges at speeds exceeding 600 cups per minute to provide a smooth, safe surface for the mouth (see Figure 3-6).

## Blisters

Blisters are custom-formed for individual products where standard rectangular or round shapes are not suitable. Thermoforming small quantities of blisters may often involve utilizing low-cost thermally inefficient molds and manually handled steel-rule dies so that tooling expenditure is minimized.

Clear-plastic blisters, after being filled with items to be packaged, are often heat-sealed to cardboard to complete the package.

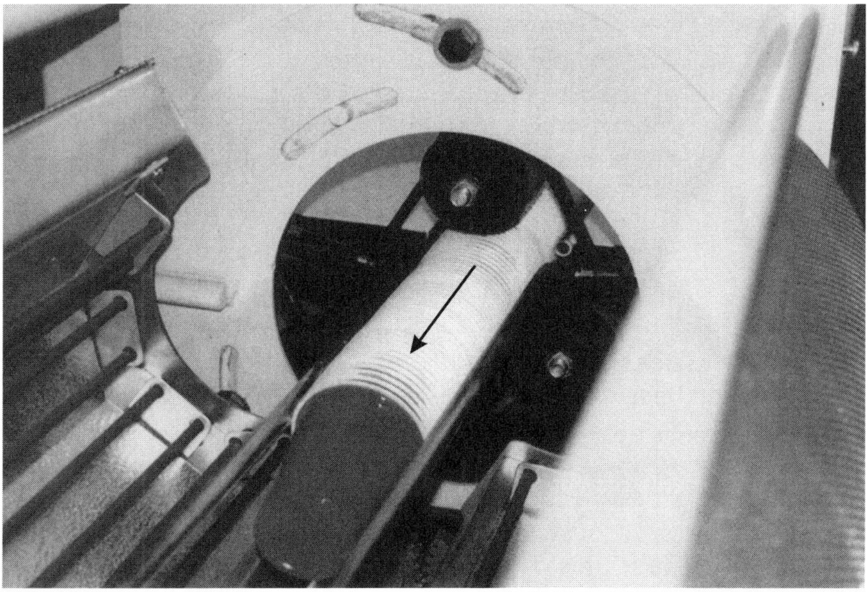

*Figure 3-6. Cup rim rolling machines roll the lips of the cups under their flanges at high speed and provide a smooth, safe surface for the mouth. (Courtesy Future Mold Corp.)*

The clamshell is a complete package whose contents can be removed and the package reused to store the remaining items. When the product is encased in a clear clamshell blister, it can be viewed from both sides, providing a marketing benefit. This type of package may be thermoformed with an integral hinge-and-snap locking arrangement, or it can be fabricated in two pieces and snap-locked together. When a product has a high value, as in the case of electronic or expensive consumer goods, the package—consisting of a PVC blister and cover—can be radio frequency (RF) heat-sealed. This style of package discourages pilferage because the sealed area must be opened with scissors.

## Tray Packaging

Tray packaging allows a variety of consumer or industrial products to be displayed individually or in multi-pack trays in conforming custom cavities. Elaborate cosmetic packaging is marketed and stored in attractively colored or flocked thermoformed trays. The packages may be decorated with hot-stamped decorative logos, information, or advertising. The molds require design ingenuity to snugly nest loosely toleranced glass and plastic bottles. When packages are thermoformed from cotton-hair flocking bonded to a plastic sheet, the flocked face should not be thermoformed against the mold surface to avoid flattening the fur-like texture.

## Foam Packaging

Foam packaging of low-cost polystyrene or more expensive polyethylene may be used to provide shock protection for fragile products. The molds consist of two matched halves which, when they are closed, contain clearance equal to the wall thickness of the finished product (see Figure 3-7). Foam molds require large draft angles and radii to prevent rupturing sidewalls during forming. Generally, vacuum or pressure forming does not play much of a role in the forming process because closing mold halves mechanically force the hot plastic sheet into the cavities.

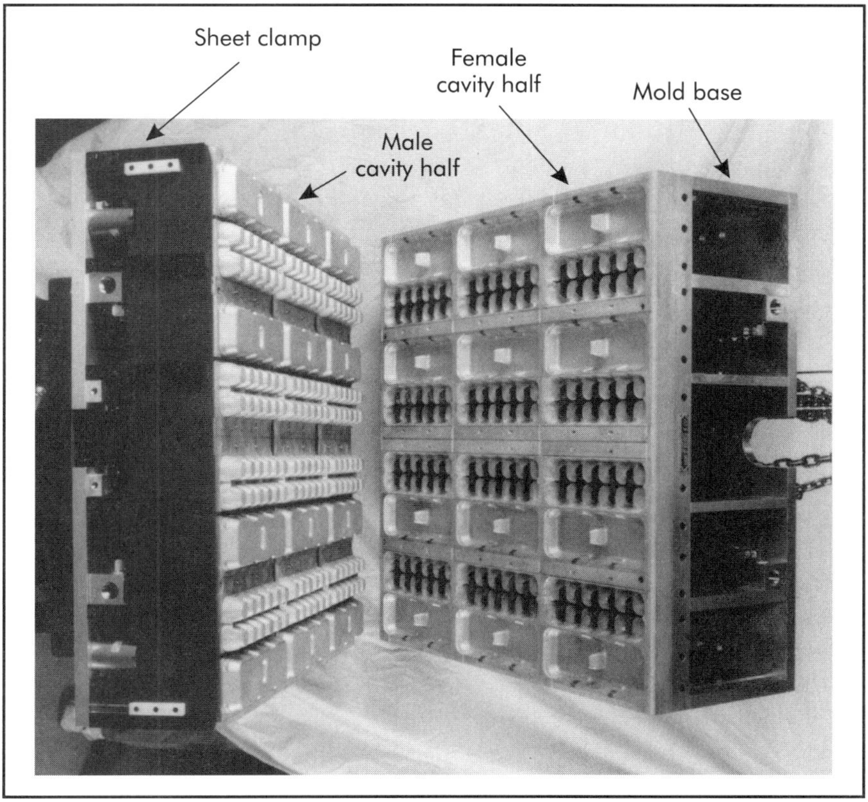


*Figure 3-7. Multi-cavity egg carton foam styrene molds consist of two matched halves which, when they are closed, contain clearance equal to wall thickness of the finished product. (Courtesy E.D. Segen & Co., LLC)*

## Continuous Formed Parts

Continuous forming of a web without trimming individual shots is specified for industrial dunnage, wastewater facility components, and products poured as liquids (jelly, butter, etc.) into thermoforming cavities and heat-sealed to a lid. Usually, a slitter is mounted onto the exit end of the thermoformer to remove excess edge trim. The web can then be wound on a large-diameter shipping core or guillotine cut to handy lengths for shipment. The mold often is designed to provide a seamless transition from shot to shot.

## Summary

The designers of thermoformed components should have knowledge of the mechanics of the thermoforming process and its tooling. Maximum sidewall draft and generous radii tend to eliminate a bevy of problems at the shop level. Maintaining reasonable trim tolerances with respect to the formed-part segment and keeping within the constraints of the equipment provide the lowest part cost.

# 4

# Thermoforming Machines

## Pioneer Developments

The two main thermoforming processes are vacuum forming, which includes skin packaging, and pressure forming.

## Vacuum Forming

E. Bowman (Bo) Stratton, pictured in Figure 4-1, was the late pioneer in commercial vacuum-forming machinery that helped develop and promote thermoforming. During World War II, he was involved with the materials development section of the U.S. Army Map Service and worked on methods to quickly produce three-dimensional sector maps rather than the usual flat representations. Stratton and his teammates heated printed plastic and vacuum formed it on a raised relief map mold. He envisioned that this process would be modified to produce advertising displays, packaging, and food containers.

Early in the 1950s, Stratton's company, AutoVac, continued to perfect the vacuum-forming process while constructing machinery and molds for the industry. Stratton traveled throughout North America talking to diverse business and technical groups and assisting novices in developing new applications for products of the infant vacuum-forming industry. Early vacuum-forming machinery firms of that period included Atlas, later named Zed Industries,

*Figure 4-1. E. Bowman (Bo) Stratton, the late pioneer in commercial vacuum-forming machinery. (Courtesy PlastiMach Corp.)*

Comet Industries, Tronomatic Corporation, and Vacform Corporation. These machinery and mold builders received feedback from customers and quickly developed many of the vacuum-forming techniques still used today.

## Pressure Forming

Pressure forming was introduced in the early 1950s to process oriented polystyrene (OPS). However, OPS proved to be unworkable in the then available vacuum-forming machines, because it required near-perfect sheet temperature control and high forming pressures. The Plax Corporation built machinery that automatically fed precut, preprinted, OPS blanks to a pressure former to make clear container lids. In the mid-1950s, Gaylord Brown designed an all-purpose, roll-fed, pressure-forming machine with a radiant oven to make plastic containers. The machine had all the elements of modern equipment.

In the mid-1950s, Robert Butzko introduced contact heat to pressure forming by using a temperature-controlled metal plate to transfer heat to the plastic sheet (see Figure 4-2). The machine was similar to the Plax Corporation process, with the additional feature of using a wrap-around, knife-like die to trim in place. This development indexed a web of plastic through a press, which formed and trimmed multiple cavities on one station and did not require precut blanks. The technique is still used today to form clear OPS food containers, salad trays, and bakery packaging at

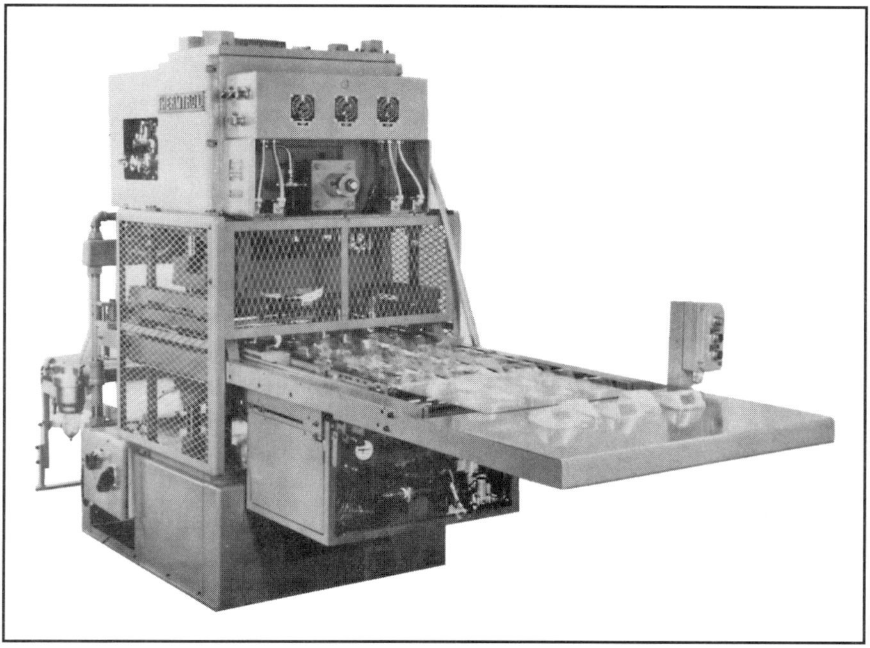

*Figure 4-2. Contact-heat pressure former. (Courtesy Thermtrol Corporation)*

high speeds in improved versions of the original Butzko design. Some of the materials that can be processed on this equipment are OPS, polyvinyl chloride (PVC), high-impact polystyrene, and polypropylene.

## Skin Packaging

Skin packaging is a form of vacuum forming in which a product is substituted for a mold (see Figure 4-3). This process requires a porous, preprinted card to which a heat-sensitive adhesive coat-

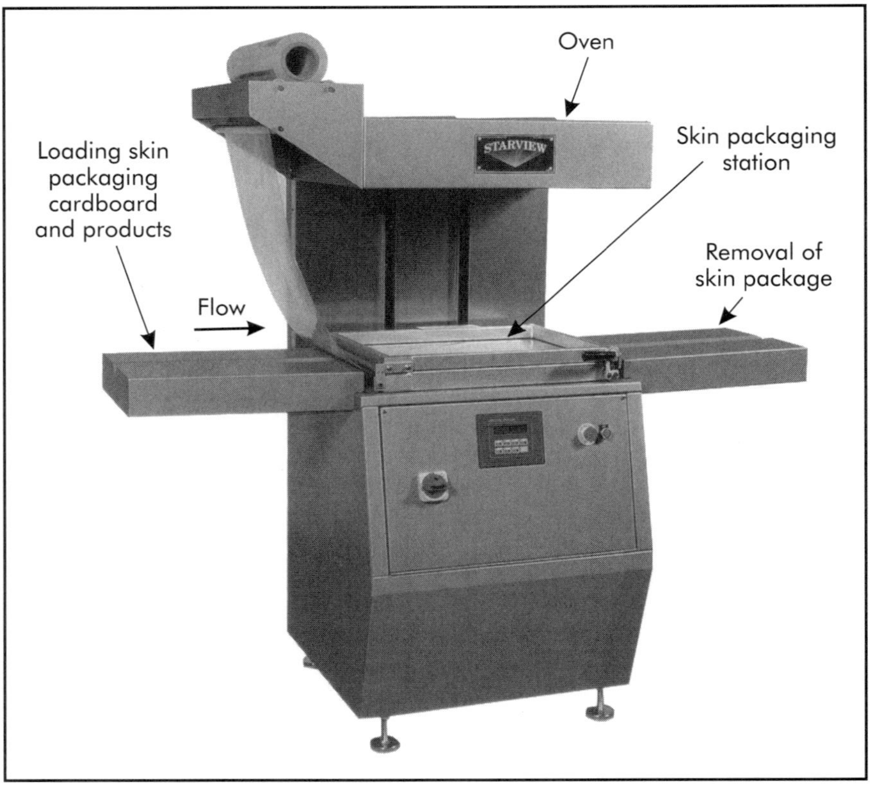


*Figure 4-3. Skin packaging is a form of vacuum forming in which a product is substituted for a mold. (Courtesy Starview Packaging Machinery, Inc.)*

ing has been applied. Products to be packaged are loaded face-up on the card. The card-and-product combination is combined with a hot-plastic web and vacuum formed (skin packaged) directly to the card. During the heating process, the adhesive becomes activated and hot plastic is drawn by vacuum against the porous substrate of the tacky card. When cool, it is cut apart by a knife-like die. Some products have geometry unsuited for skin packaging and finished packages may contain webs, thin spots, and other defects. The cost of the package is very low and users often are reluctant to switch to the more elegant heat-sealed blister package, which requires a mold cost for each product.

Skin packaging is generally considered a packaging operation best performed at the location where the product is manufactured. No molds or thermoforming skills are required.

## Roll-fed Thermoformers

Modern roll-fed thermoforming equipment capable of either vacuum or pressure forming consists of an index device driving a plastic web through several processing segments. The web is indexed through a multi-stage oven, a thermoforming press containing molds, and optional stations, which may include a trim press, part stacker, guillotine shear, or scrap grinder (see Figure 4-4).

### Indexing

Indexing the web through the machine usually, but not always, requires both edges of the sheet to be impaled onto a spike-like chain that then transports the web through the thermoforming machine (see Figure 4-5). The chain index drive can be a mechanical device (rack and gear) or, for increased accuracy, an electrical servomotor drive (see Figure 4-6). Index accuracy is of great importance on all in-line form-and-trim machines where good registration of the formed part to the cut edge is essential.

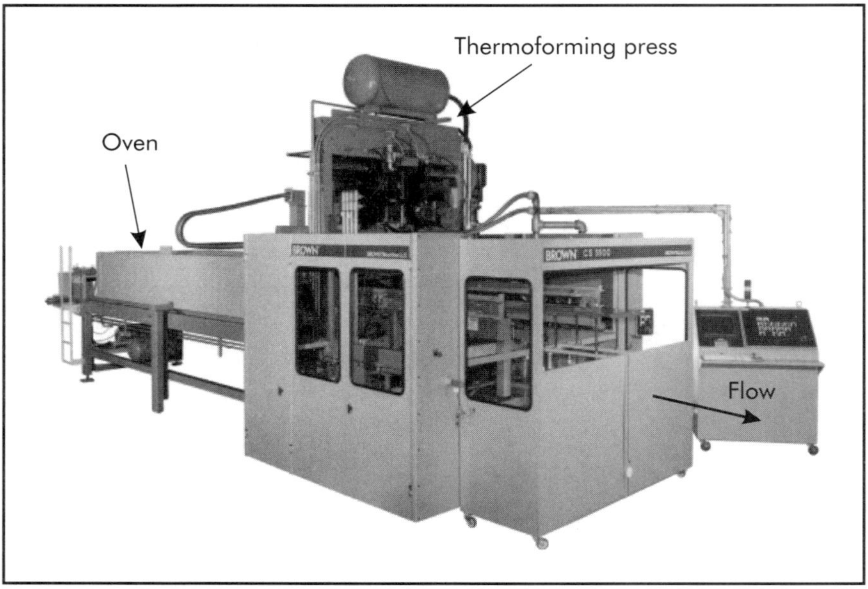


*Figure 4-4. In a modern thermoformer, the web is indexed through a multi-stage oven to the thermoforming press. (Courtesy Brown Machine Co., LLC)*

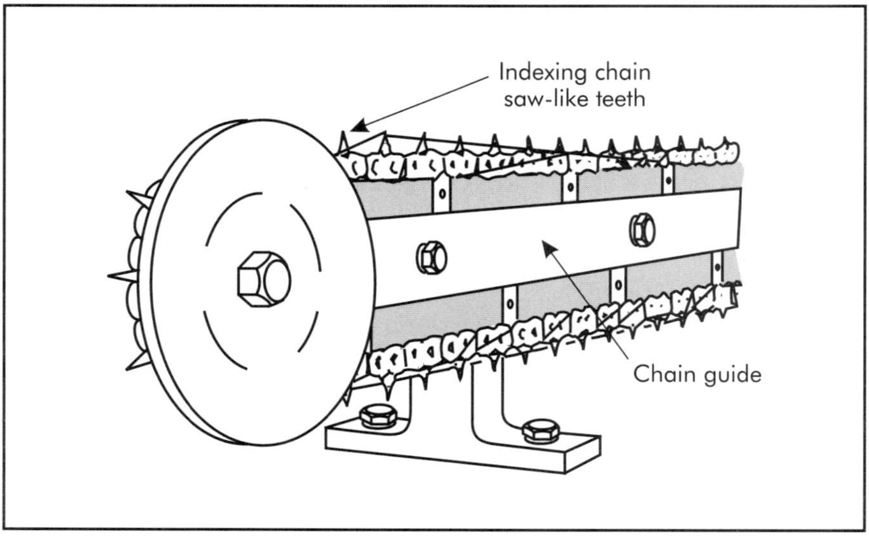


*Figure 4-5. Indexing the web through the machine requires both edges of the sheet to be impaled onto a spike-like chain that then transports the web through the thermoforming machine (Mitchell 1998).*

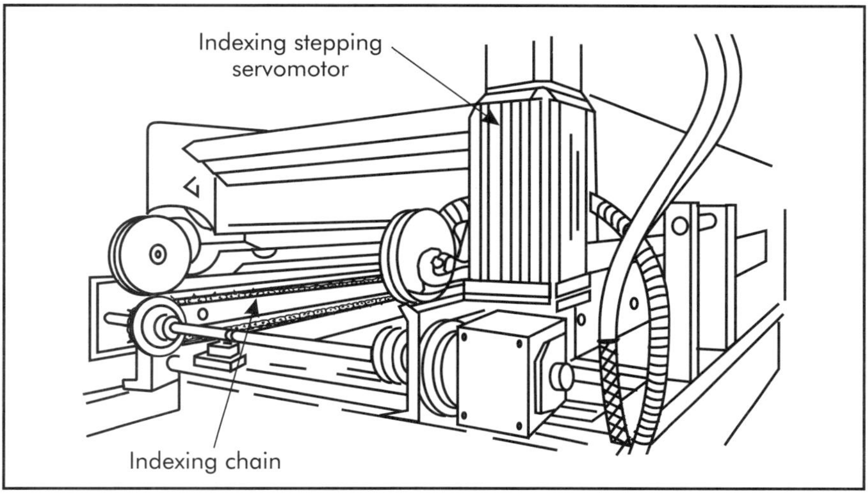


*Figure 4-6. Indexing-chain electrical servomotor drive (Mitchell 1998).*

## Oven Heat Sources

The time required for heating the web is zero because ovens preheat the plastic web to forming temperature while the previous shot is being thermoformed in the mold. A mold extracts heat from the thermoformed plastic and discharges heat to cooling liquids circulating through the mold. Overall, machine cycle time is based on the cooling efficiency of the mold, plus index motion timing between stations. Thermoforming ovens are equipped with instrument-controlled heating elements, usually with two or more index stations and with upper and lower banks of heating elements. When the machine is stopped for any reason, the ovens are removed from the web because residual heat may cause the plastic web to melt or burn. Oven removal can be accomplished manually or automatically, either by swinging open the top and bottom oven sections or by moving the entire oven assembly out parallel to the web (see Figure 4-7).

There are many types of heating sources available to provide controlled heat for the oven. Some of them are:

- Tubular heating elements, which can be designed to go across or parallel to the indexing web of plastic. These elements can

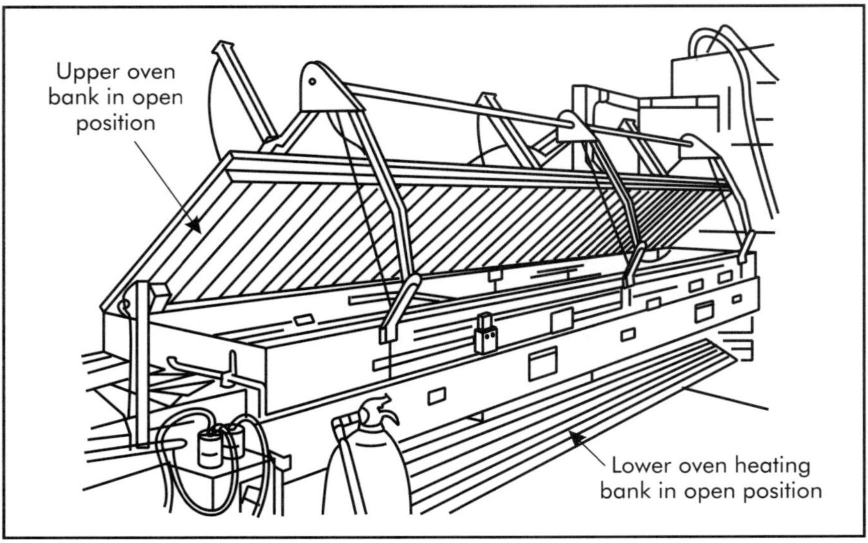


*Figure 4-7. Oven opening can be accomplished manually or automatically (Mitchell 1998).*

be bent to shape, if needed, and watt density can be varied to suit requirements for uniform web temperature.

- Radiant heating panels of various designs, which are attempts to provide an efficient black-body energy emitter (see Figure 4-8). Some of these units use an emitter 1 in. (25.4 mm) wide bundled into an assembled panel, with many panels making up an oven section. Similar assemblies are specified using quartz or ceramic glass panels.
- Ceramic heating elements, which are relatively small individual units mounted in a pattern that can be controlled in zones (see Figure 4-9). This system allows very close temperature control of any given area of plastic web. However, the availability of too many individual control adjustments can cause confusion to machinery setup personnel, as the ovens take 15–30 minutes to reach thermal equilibrium conditions for a steady state of consistent sheet temperature. Computer control of oven heat for each mold is useful when temperature adjustments are known and can be stored in the computer memory after settings have been established by the initial production run.

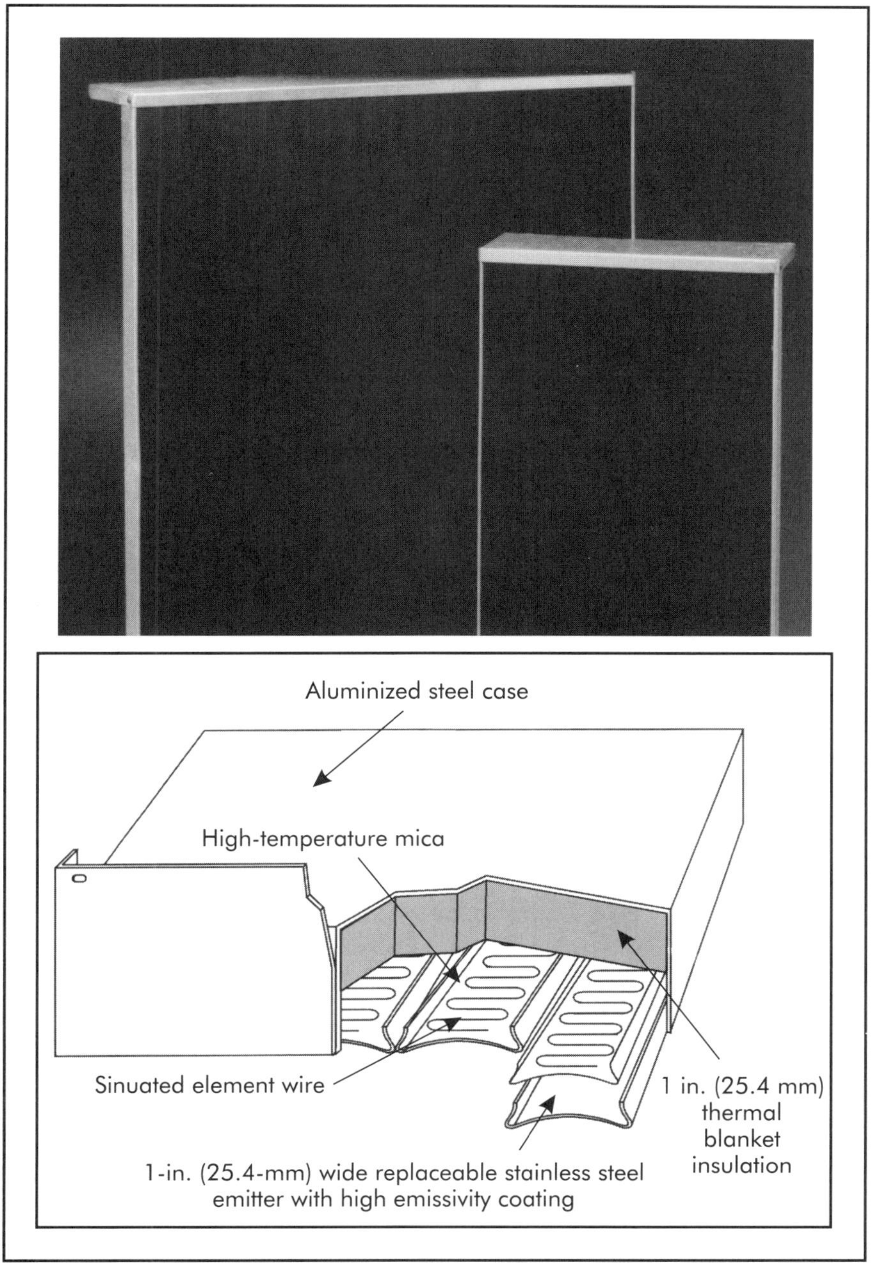


*Figure 4-8. Radiant heating panels. (Courtesy Watlow Electric Manufacturing Co.)*

(a)

(b)

*Figure 4-9. (a) Ceramic heating element (b) ceramic heating panel. (Courtesy Infrared International)*

- Catalytic gas heaters, which derive their thermal energy from the reaction of oxygen (from air) and natural gas to a hot platinum catalyst built into a panel. These panels never burn the gas, but the reaction, which is flameless, produces heat, carbon-dioxide gas, and moisture vapor. Manufacturers claim uniform temperature and more economical operations than those systems using electrical energy.

Heat loss inside the oven to metal chains and indexing rails could cause the edges of the plastic shot to be lower in temperature than the rest of the plastic web. Side-rail heaters installed above indexing rails increase the temperature of sheet edges so the entire shot can be heated uniformly. As the sheet indexes out of the oven through the opening in the oven enclosure, the open space may cause the leading edge of hot sheet to cool down. The oven exit may require additional heater banks to replace heat loss caused by this opening.

## Thermoforming Presses

Thermoforming presses are designed to allow the mold to be mounted on either the upper or lower platen with vacuum- and pressure-forming utilities available nearby. The maximum retract-to-extend stroke of press platens averages 6–8 in. (152.4–203.2 mm), plus a tool buildup space on each platen. On some machines, the shut height of the top platen may be mechanically adjustable over a short range, which allows the mold halves to close tightly together for pressure forming. When this feature is not available, tooling can be shimmed to the correct shut height.

### Lockup Force

Pressure-forming requires that the press have sufficient lockup force to resist the internal opening force of the forming pressure. This force lockup can be calculated by:

$$F = AP \quad (4\text{-}1)$$

where:

$F$ = force to resist tool opening, lb (N)
$A$ = area of mold, in.$^2$ (cm$^2$)
$P$ = pressure-forming air pressure, psi (Pa)

Presses are rated for a maximum mold area at a specific maximum forming pressure, or in tons of lockup force. If pressure or mold area is exceeded, the force of the internal mold opening vents the excess forming air to the atmosphere.

## Toggles

Many presses utilize toggle mechanisms to quickly close and lock platens for forming (see Figure 4-10). The closer together that the toggle links are extended into a straight line, the greater their locking force. If pins and bushings of the toggles are worn, locking force can be seriously diminished as toggles bend away from a straight line under pressure-forming load. Maximum weight per platen of a tool may be restricted in a toggle press. This can occur as the angle between links and force of the driver (cylinders or electric motors) determine lifting force at the maximum retracted position. Presses may require a balance cylinder to coun-

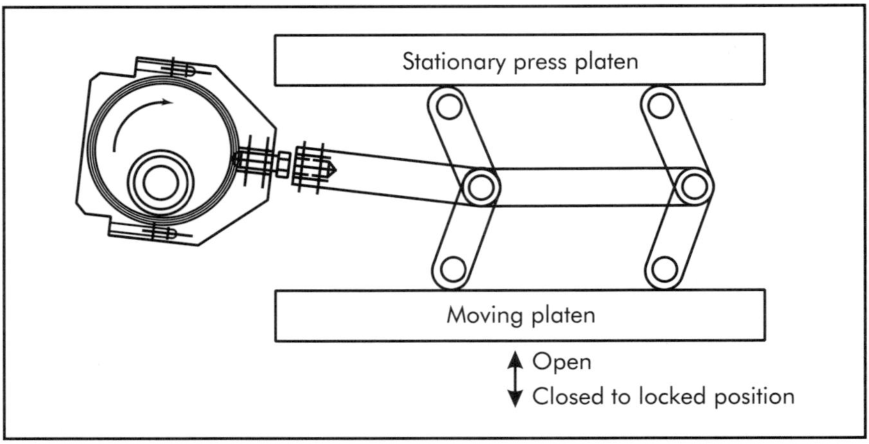


*Figure 4-10. Many presses utilize a toggle mechanism driven by a servomotor to quickly close and lock the platens for forming (Mitchell 1998).*

terweight the rated tool weight. Platen toggles driven by air are inherently variable speed devices due to the geometry of linkage design and the inability to continuously vary the speed of the cylinder rod while operating. Hydraulic cylinders can operate at constant velocity, but air is a gas, which is compressible and can not be as easily controlled as a liquid. Servomotor activators directed by a programmable processor allow the platen to operate at a variable accelerating or decelerating velocity, which results in a controlled stroke throughout its length. Controlled velocity during platen closing while the plug distributes plastic within a cavity is an important variable for obtaining a uniform wall thickness in the formed part. Servomotor speed control of the platen closing also prevents impact damage to the mold when locking tooling.

## Control Systems

Process-control systems supplied with modern thermoformers are most often computer controlled with a visual panel display. Many early model thermoforming machines built before 1980 were designed with electric timers and relays hard-wired together to operate the forming cycle. The basic forming cycle steps remain much the same on both types of control systems.

The basic thermoforming cycle is:

1. Index the heated plastic web into the thermoforming-press mold area.
2. Close platens and choose which platen is to be extended first.
3. Vent atmospheric air out of the cavity, sometimes using a partial vacuum as platens are closing. This vacuum may be necessary to remove the air out of the cavity as the plug travels within the cavity.
4. Vacuum or pressure forming occurs after platens are locked.
5. Exhaust pressure-forming air from closed mold before retracting platens.
6. Open platens, choosing which platen is to be retracted first.
7. Eject formed shot mechanically or with blow-off air pressure.
8. Platens stay open to clear tooling while indexing formed shot out of the press.
9. Recycle.

## Prototype Parts and Machines

Thermoformed prototype parts should be produced before building a multi-cavity tool because they assist the customer and tool designer in achieving joint objectives. Often the customer provides a product drawing or a proposed part as a starting point for tool design. The part design, when thermoformed as a prototype, can be immediately approved, disapproved or, more likely, modified to increase its usefulness. This is a case where one prototype part is worth a thousand product drawings, requiring the rework of only one cavity rather than a multi-cavity production mold.

Prototype thermoforming machines, as shown in Figure 4-11, should have the same process capabilities as a production machine or resultant parts may not be of the same quality. Thermoforming techniques used to make sample parts should mirror the production process. Tooling components that control the uniformity of wall thickness, such as plugs or assists, may require modification of their geometry during the sample-making process to produce the desired properties. Modified dimensions then can be updated by the tool designer on finished production-mold drawings and used for final tool design.

Some prototype forming machines in use do not have pressure-forming capability, so their output may not correlate directly to the pressure-formed parts' final detail. Vacuum-formed prototypes made nearly the same as production pressure-formed parts can be a disaster if supplied to the customer with a promise that the production parts will be much better quality. This is not to say that a highly skilled vacuum-forming prototype maker cannot produce high-quality uniform sample parts. However, pressure forming is more forgiving for less-qualified operators. Some modern well-equipped thermoforming shops consign their antique (junk) vacuum-forming equipment to the sample-making department, where many prototypes do not come up to standard.

## Vacuum-forming Machines

Vacuum-forming roll-fed machines are built and sold for limited volume work and, often, are provided with the most rudimentary

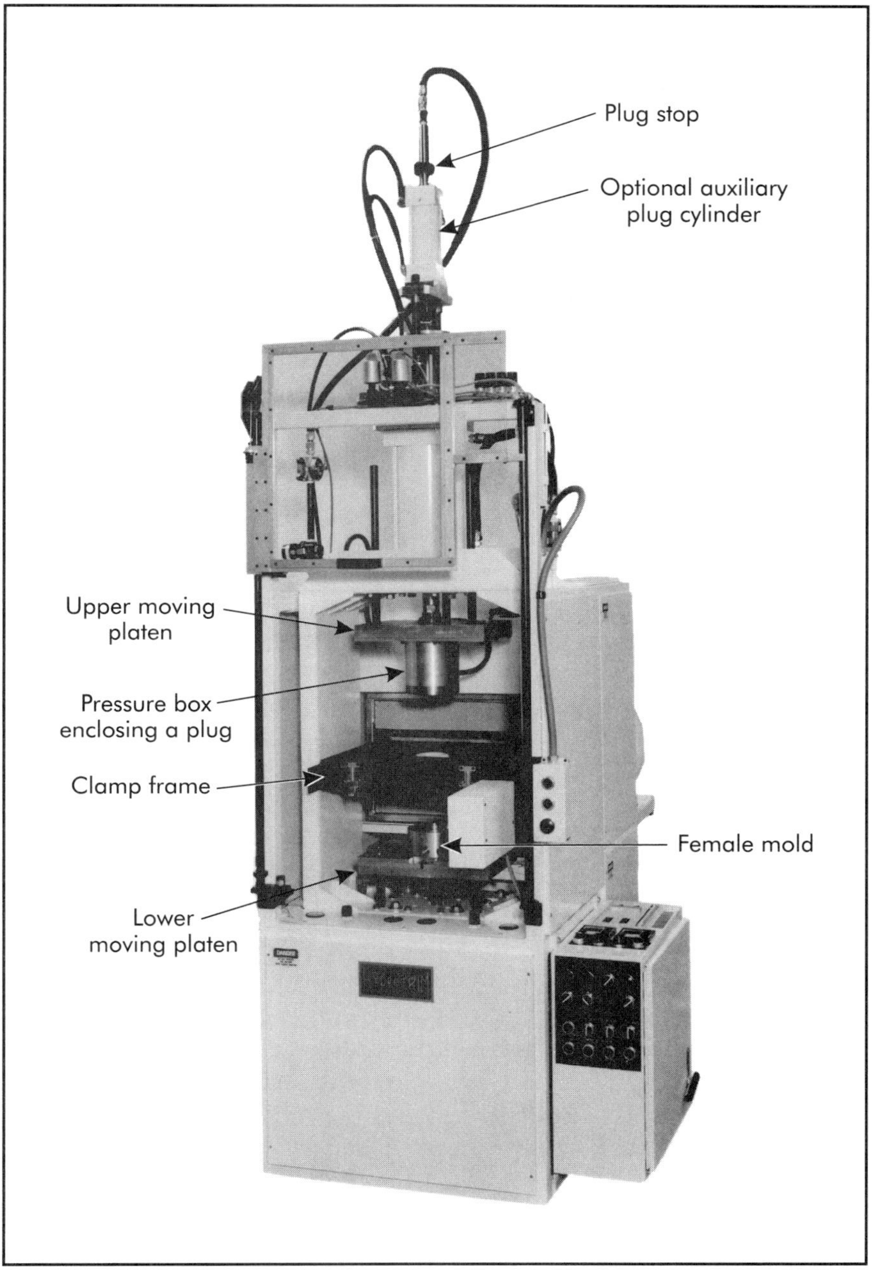


*Figure 4-11. Prototype thermoforming machine. (Courtesy PlastiMach Corporation)*

molds (see Figure 4-12). This type of equipment is relatively inexpensive and does not require sophisticated controls, dead-accurate indexing, or high-tonnage pressure-forming presses. The production rate of vacuum forming can be 150% slower than pressure forming. Pressure forming drives the heated plastic into more intimate and efficient contact with the cavity than vacuum forming. The heat is then transmitted from the cavity to the mold base, where it is carried away by temperature-controlled liquids. Since the quantity of vacuum-formed parts is often small, mold heat transfer can be simpler and less costly than that in a corresponding pressure-forming tool. The completed vacuum-formed shots are guillotine cut and trimmed off-line manually at greater cost than a continuous in-line trimming process. The ovens supplied with this machinery often are less sophisticated in design and may require a great degree of skill to maintain a uniform heating pattern on the plastic web.

## Trim Presses

There are two types of trim presses: on-line and off-line.

### On-line Press

An on-line trim press built on a thermoforming machine base cuts formed parts using a knife-like die then indexes the shot to an optional, in-line automatic-stacking press or guillotine shear. The trimmed parts are lightly held in the shot by nicks for subsequent manual or automatic stripping (see Figure 4-13). This trim press can also accommodate a punch and die set, if provision is made for removal of trimmed loose parts from the press area. When two trim stations are necessary to complete the finished product, the first-stage trim can be accomplished on the on-line trim press and the completed trim on an off-line trim press. This type of online equipment is most useful for short production runs of 500–10,000 shots because the knife-like dies provide trimming at no extra labor cost. Additional auxiliary operations, such as cementing or heat sealing of a backing cardboard, can be accommodated

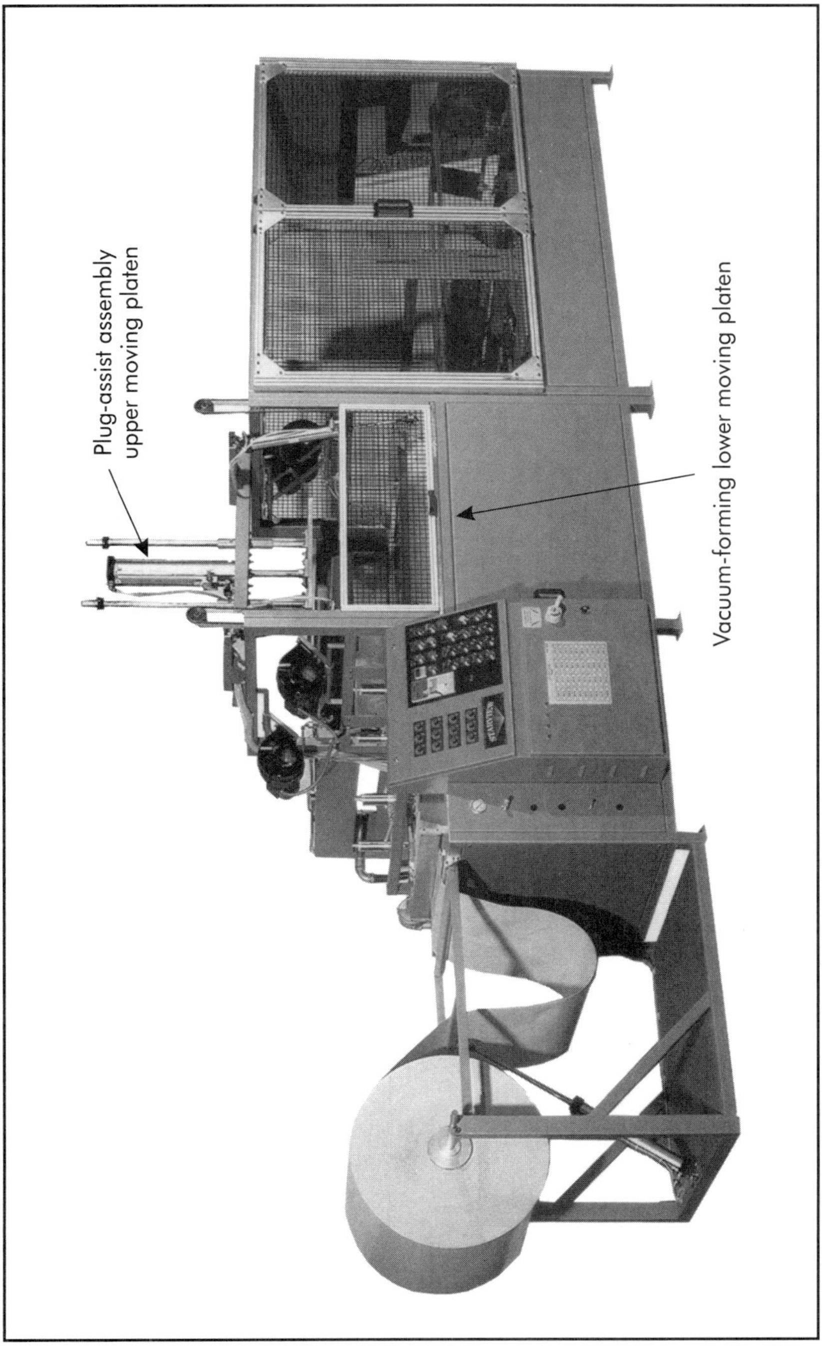


*Figure 4-12. Vacuum-forming roll-fed machine. (Courtesy Starview Packaging Machinery, Inc.)*

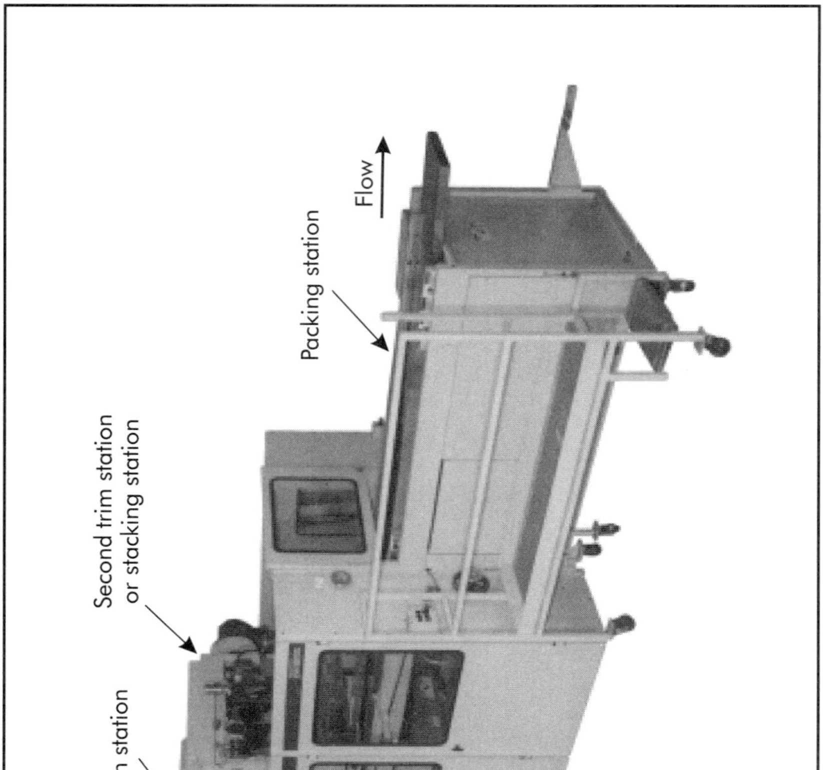


*Figure 4-13. In-line thermoforming, trim, and packing machine. (Courtesy Brown Machine Co., LLC)*

on this machine. Cost of the equipment is in the medium range. These machines are the workhorses of the custom thermoforming industry and many in-plant thermoforming operations.

## Off-line Press

An off-line trim press, run in tandem with a thermoformer, usually trims parts with a punch and die, but some knife-like die-cutting machines are available for this purpose (see Figure 4-14). All off-line trim presses have their own integral indexing systems to operate free of the thermoformer's index system. A horizontal trim press with a punch and die trims parts so that they stack naturally, moving out in the horizontal direction where they are easily counted and packed for shipment. These *horizontal trim presses* are so named because their power stroke moves in this direction. *Vertical trim presses* direct their power stroke perpendicular to the floor and use both knife-like dies and punches and dies. Both types of trim presses can be used in tandem with an on-line thermoformer for progressive trimming of parts requiring two dies. An off-line trimming system operated with punch and die requires the largest investment, including the additional amortization expense of the trim press.

## Cut-in-place Contact Heat Thermoformer

The cut-in-place, contact-heat thermoformer is the only style of thermoformer that heats plastic by thermal conduction rather than radiant or convection heat transfer (see Figure 4-15). The cut-in-place thermoformer contains a heated plate with its temperature closely controlled and set at the softening point of the plastic. Attached to the hot plate is a hardened-steel blow plate. The blow plate is ground true and parallel with small drilled holes, 0.020 in. (0.51 mm) in diameter, connected by a manifold to the thermoforming air valves. When the press closes on the web of plastic, air is introduced through the mold, forcing the cold plastic into close contact with the heated blow plate (see Figure 4-16). On the face of the plastic touching the hot plate, entrapped air is

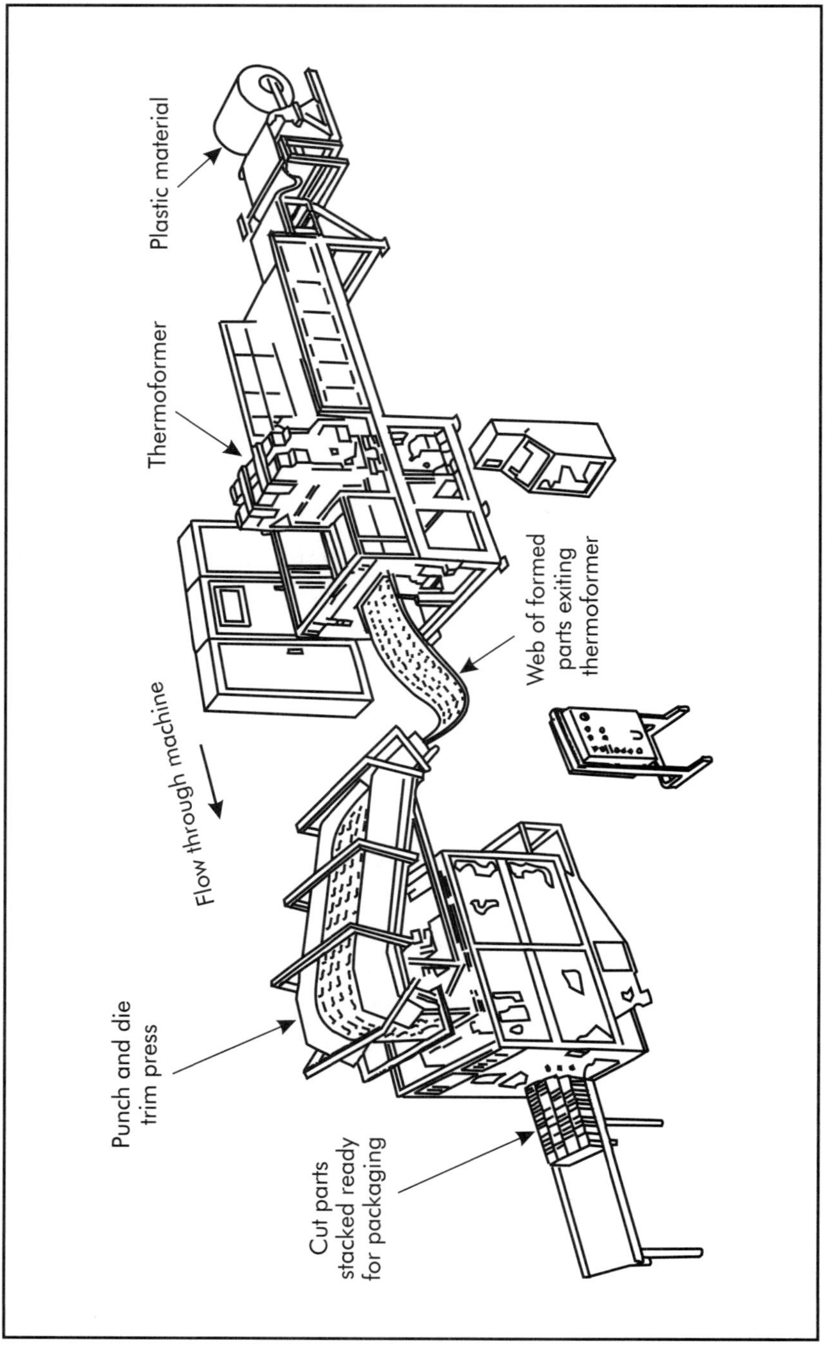


*Figure 4-14. An off-line trim press run in tandem with a thermoformer (Mitchell 1998).*

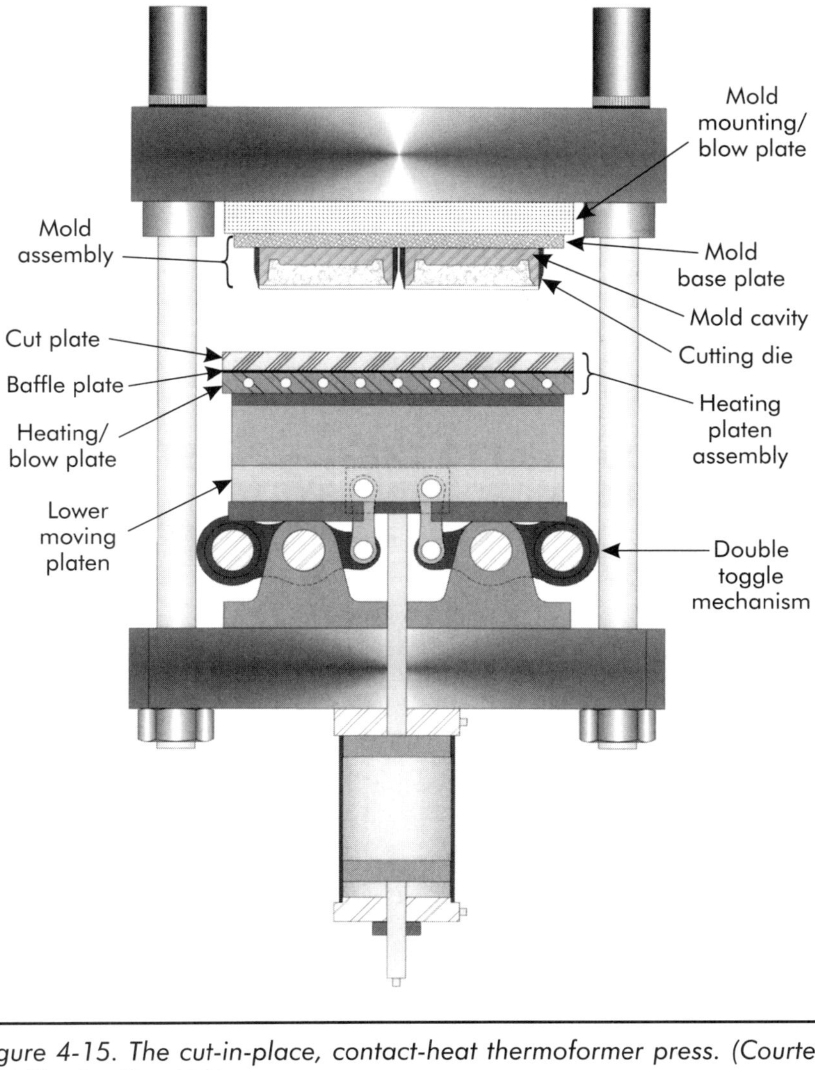


*Figure 4-15. The cut-in-place, contact-heat thermoformer press. (Courtesy GN Plastics Co., Ltd.)*

vented through tiny holes in the blow plate. The process is reversed after the plastic is thoroughly heated and forming pressure air is directed to the plastic through the blow plate holes. This forces the plastic to take the shape of the mold. When the

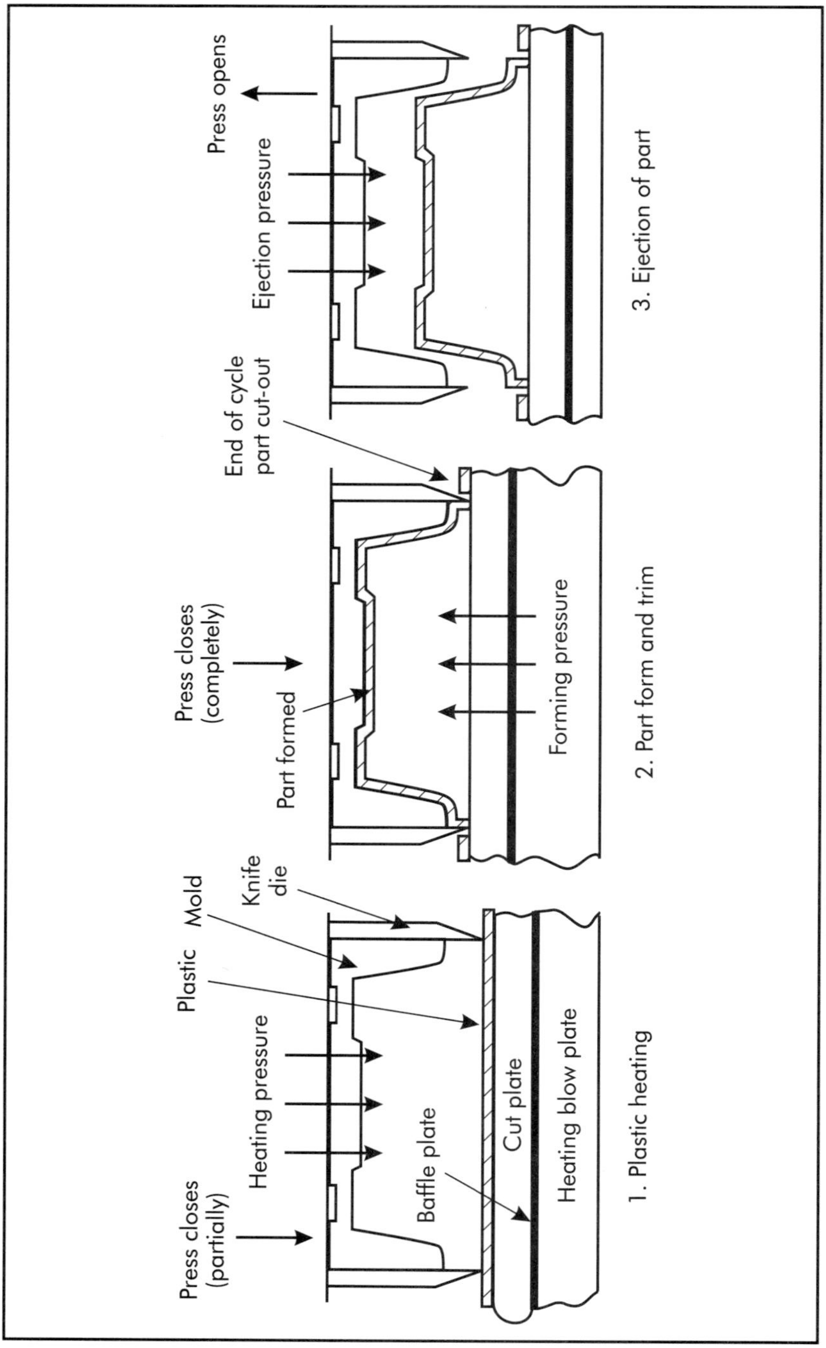


*Figure 4-16. Contact heat thermoforming process (Mitchell 1998).*

plastic is cooled in the mold, further closing of the press platen trims the finished part when it meets the knife-like die that encircles the mold. Finished multiple parts are held in place in the web by designing slight imperfections in the cut edge so that cut formed parts are indexed out of the press as a single shot. When the web is clear of the press, it is bent 90° over a roll. Trim imperfections break, freeing finished parts, which are directed to a stacking device. The web is then wound on a mandrel for recycling.

This type of cut-in-place thermoforming uses the most exact method to heat the plastic since the large heavy mass of the heated plate maintains a uniform temperature. Oriented polystyrene sheet (OPS), which is difficult to thermoform, can be heated to a consistent specified forming temperature, thus easing production and quality problems. Trimming is very accurate and does not require a separate trim press because the inexpensive knife-like die is wrapped around the perimeter of the mold. Disadvantages of this technique should also be recognized:

- Only female molds are adaptable to the process since the mold cannot extend above the die.
- Parts that have heavy undercuts may be difficult to eject, since the whole shot with its formed, trimmed components loosely held to the web must be indexed out as a unit.
- The process does not lend itself to use of auxiliaries, such as plugs, to uniformly distribute material.
- The plastic web is usually heated from one side only, limiting the practical thickness of plastic web that is to be processed rapidly. The maximum practical sheet thickness is 0.040–0.060 in. (1.02–1.52 mm), since in this process the time to heat the plastic is additive to the overall cycle, although auxiliary preheating ovens help reduce the cycle.
- Some clear plastic sheets other than oriented polystyrene may not be formed water-glass clear, since the contact-heated plate imprints a slight haze on the face of the sheet.

A very large industry flourishes using contact cut-in-place thermoforming to produce food and takeout containers, since the equipment and these molds are inexpensive and the production technique is easy to master.

## Trim-in-place Preheat Thermoformer

The trim-in-place preheat thermoformer is the most common machine to be found outside of North America and it produces parts with the greatest accuracy of trim relative to thermoformed area (see Figure 4-17). The oven and index arrangement is similar to other roll-fed machines, but tooling and the forming press are markedly different. The punch and dies are combined with the mold into one tool, so that thermoformed parts are trimmed free from the web in the press section and ejected out as individual parts. The latest machines allow the lower platen of the press in the open position to rotate to an angle of approximately 30–45° at the end of the trim cycle. The cut parts are then moved out of the die section into collating chutes for packing. The lower platen then rotates to the operating position to recycle the machine. This type of mechanism requires dies and punches of highest precision and quality.

## Rotary-drum Mold Thermoformer

Hot-sheet extrusion onto a rotary-drum mold thermoformer is a continuous, rather than intermittent, thermoforming process. It is quite different from previously mentioned equipment. There is no need to reheat the sheet when using this equipment because the extruder continually presents hot plastic to a rotary drum mold (see Figure 4-18). The heated sheet is moving along to synchronize with the rotating mold cavities. The sheet is thermoformed as the drum revolves and a fresh series of empty cavities is made available to the extruded plastic. As the drum rotates, the mold cools the thermoformed parts and the finished line of parts moves out of the machine to be trimmed on standard off-line presses. Not all types of parts can be produced, but it is a very efficient process for suitable items, such as lids and shallow, female-thermoformed parts. The mold cost is greater than for a standard tool since cooling water, vacuum, and air must be piped into a rotating drum and distributed to the molds.

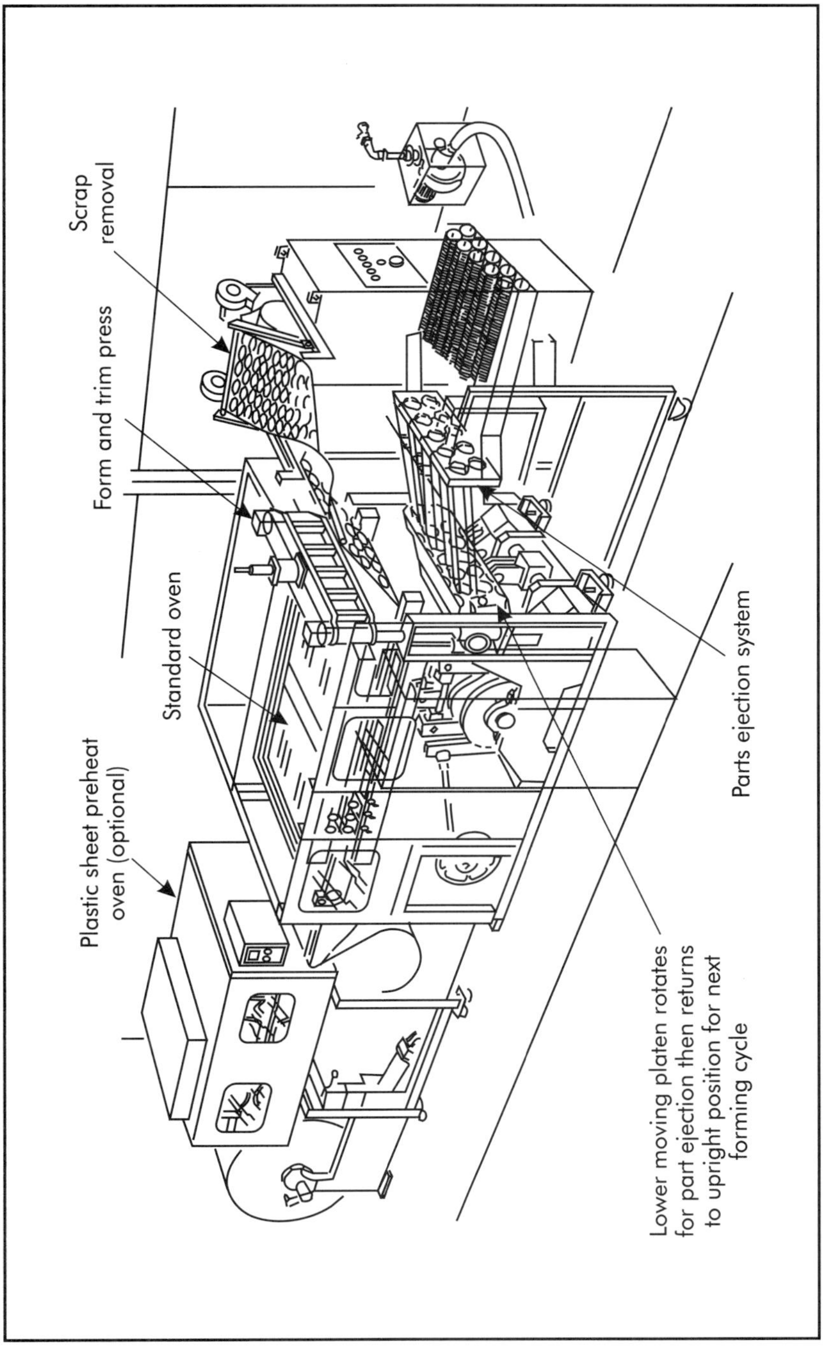


*Figure 4-17. The trim-in-place preheat thermoformer (Mitchell 1998).*

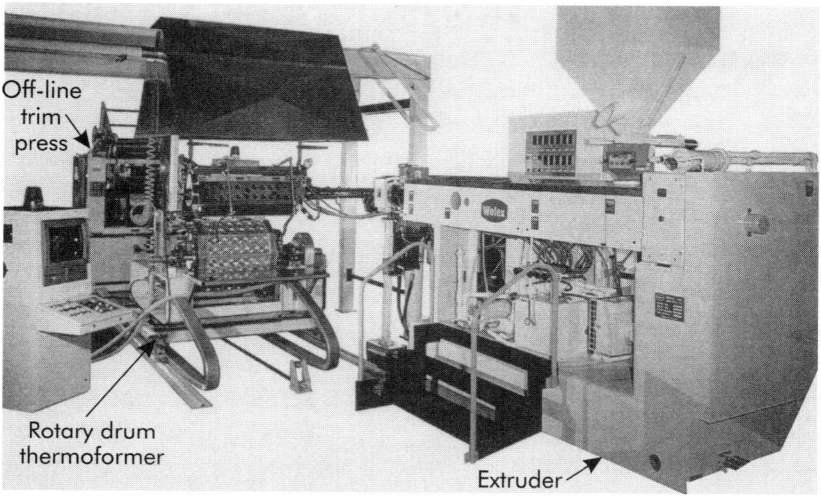


(a)

(b)

*Figure 4-18. The extruder continually presents hot plastic to a rotary drum mold. (Courtesy International Thermoforming Systems)*

## Working Together

The thermoforming engineer should try to remain in constant contact with the machine operators of various types of thermoforming machines to be informed of any serious production problems. Many glitches that hinder production are not properly communicated up the corporate ladder to an engineer who has the ability to solve these problems. By venturing into the production area, management interacts more informally with machine operators. An engineer demonstrates an interest in being part of the team just by visiting the production floor. This encourages suggestions vital for improvement of molds, dies, and equipment.

## Reference

Mitchell, Philip, ed. 1998. *Tool and Manufacturing Engineers Handbook*, Fourth Edition, *Volume 9: Material and Part Handling in Manufacturing*. Dearborn, MI: Society of Manufacturing Engineers.

## Pioneering Founders of Thermoforming Machinery Firms

Atlas: later named Zed Industries: Vandallia, Ohio. David Zelnick

AutoVac Corporation: Bridgeport, Connecticut. E. Bowman (Bo) Stratton and his partner, engineer Robert Butzko

Brown Machinery Corporation: Beaverton, Michigan. Gaylord Brown

Comet Industries: Chicago, Illinois. The Kostur family

Plax Corporation: Bloomfield, Connecticut

Thermtrol Corporation: Bridgeport, Connecticut. Robert Butzko

Tronomatic Corporation: New York. Joseph Swick

Vacform Corporation: Long Island, New York. Sanford Zimmerman

# 5

# Trim Presses

Individual thermoformed parts can not be used until they are cut from the plastic web with trim equipment and dies. Early on, the thermoforming industry adapted trimming machinery from the printing, metal, and leather industries.

Printers and cardboard box makers rely on steel-rule dies because they are low cost, accurate, and can be delivered rapidly. A highly developed network of steel-rule cutting machinery, techniques, and local die sources existed prior to the invention of thermoforming. Sheet metal fabrication requiring a punch and die is a sophisticated art that has flourished since the industrial revolution. Shoe and leather goods manufacturers used long-life, knife-like forged dies in swing-arm mechanical and hydraulic-clicker presses 100 years prior to the advent of thermoforming (see Figure 5-1).

Early thermoforming production engineers often had previous experience in established industries and transplanted the existing trim arts to thermoforming. Unfortunately, paper, cardboard, leather, and metal materials are flat, and the existing trimming machines often did not have the necessary clearance to allow entry of thermoformed parts, which have much higher profiles. Eventually, machinery was modified and then specifically designed to process a thermoformed web. Early applications for thermoforming included the packaging of consumer products requiring clear blisters and cardboard-box thermoformed inserts. The quantities of these components purchased in the beginning were small: 1,000–50,000 parts requiring low-cost molds and inexpensive steel-rule dies. These orders were thermoformed and the shots were cut off

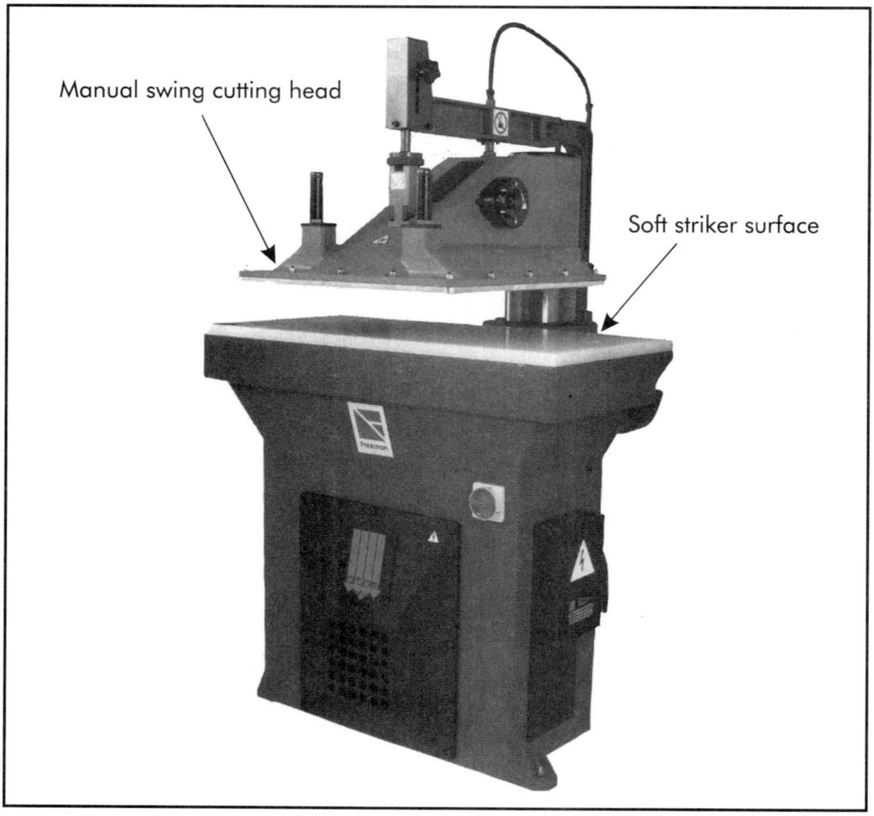


*Figure 5-1. Swing-arm clicker manual die cutter. (Courtesy L.G. Freeman Co.)*

the web. Then, the shots were manually die-cut into individual parts in clicker-type or hydraulic die cutters.

## In-line Trim Press

Modern machinery for in-line, knife-like die cutting of thermoformed products should have the following characteristics:

- A large bed size capable of handling a complete individual thermoformed shot, because the whole unit is trimmed at one time.

- Sufficient tonnage to trim multi-cavity dies at a force of 100–300 lb/in. (17.5–52.6 N/mm), of die length, with the higher-end figure used for curved dies.
- A provision for the cutting platen stroke to be controlled so that the sharp edge of the knife-like die is not dulled by cutting through the plastic and digging into the striker plate.
- The press platens or tooling buildup should not have any meaningful deflection under the trimming load. If substantial deflection takes place under the load, a die would be unable to cut through the plastic because any additional force would increase the amount of deflection and prevent trimming.
- The knife-like die cutting requires flat, straight, and smooth platens; and
- Safety and common sense dictate that all presses are potential accident hot spots, so suitable guards, electrical fail-safe controls, and employee safe-operation procedures should be taught, monitored, and enforced.

## Punch and Die Press

An off-line punch-and-die press for thermoforming should possess the following features:

- The press-bed area should be as wide as the maximum shot width and, since only one or two rows are trimmed at one time, it should be long enough to properly support the die set.
- The cycle speed of the press must accommodate the output of the thermoformer. For example, if a mold is designed with either two or eight rows of cavities per shot and thermoforms shots at 12 cycles per minute, the trim-press speed to cut out all the parts formed by the thermoformer can be illustrated with the following:

  2 trim rows × 12 thermoforming cycles/min
  = 24 trim strokes/min

  8 trim rows × 12 thermoforming cycles/min
  = 96 trim strokes/min

The trim press must be equipped with a variable-speed drive or an intermittent clutch to handle the great variation in speed needed to coordinate with thermoformer output.

The amount of force required to trim plastic using a punch and die is less than that required for a knife-like die. The force required is dependent on punch and die clearance, length of die perimeter, spring force built into the stripper plate, plastic thickness, and shear strength of the plastic. Press tonnage of 20–40 tons (178–356 kN) is common and dependent on flywheel speed and angle of crank to the moving platen at the moment of cutting.

Adjustable-stroke presses are a necessity because a shallow part will not require as much press opening space as a deeper part and it can be run at a higher cyclic rate. The higher strokes per minute thus made available for trimming a shallow part may avoid the need to provide additional rows of dies to keep pace with thermoformer output.

Deeper parts need additional clearance while being indexed into the die area, and the press crank stroke can be manually readjusted to compensate for the difference in part height. The greater the press opening, the more limited is the allowable stroke rate of the press due to vibration and high acceleration force induced into the moving platen.

The platen exit opening dictates the maximum part length that can be trimmed. Most presses have an exit opening at least 12 in. (304.8 mm) wide with somewhat larger openings available on other model machines.

Press indexing systems are synchronized to the press stroke. Web movement into the die can commence once the press opens sufficiently to clear the part. The indexing web pusher must be timed to retract fully out of the die as it closes or it can cause damage to the tool. If the time interval of the forward index cycle is not sufficient, the web may accelerate at too rapid a rate and plastic parts being indexed can be damaged. This problem of high web acceleration occurs when the cyclic rate is too rapid for the index distance and high pusher forces build up. Reducing the thermoformer output cycle might be the only way to end this troublesome condition.

## Manually Fed Trim Press

Manually fed trim presses can process individual shots that have been previously guillotine cut from a roll-fed thermoformer (see Figure 5-2). Manual handling of shots is the most expensive method of trimming and production is very slow, 1–3 shots per minute. Roll-fed thermoformers form 6–20 shots per minute and the only human intervention needed is to supply rolls of plastic and check finished-part quality. The thermoformer produces individual shots more quickly than they can be manually trimmed due to this inequality in process time. This surplus of shots builds up and requires valuable floor space for storage. Shots may become dusty and dirty while awaiting trimming.

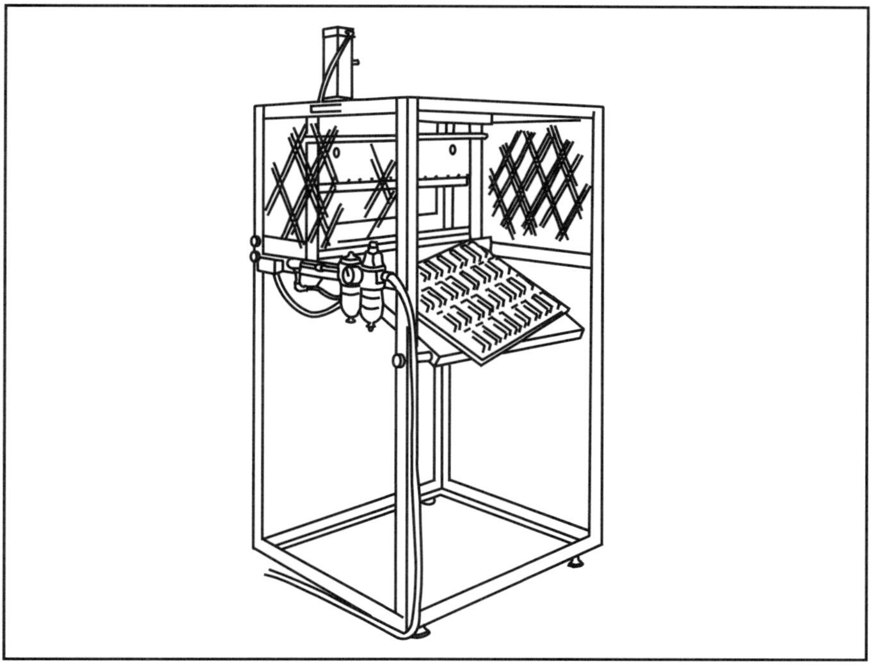

*Figure 5-2. Guillotine shear trims shots at end of thermoforming line (Mitchell 1998).*

Some of the reasons for manually trimming shots rather than using the more efficient on-line trim system are:

- Machine operators often may not have the skills to quickly set up a short-run, on-line, automatic trim press. The scrap rate and downtime in this case often is too costly to tie up an on-line thermoformer for a short run.
- The on-line thermoformer may be equipped with an index mechanism that is unable to achieve the accuracy of index required for a specific job. The die shrinkage coefficient for a still-hot shot indexing between a thermoformer and a trim press station may not be correctly calculated for an accurate register to the die. On the other hand, an individual shot left to stand until cold and stable can be calculated and a die built to operate properly in a manually fed die cutter.
- An on-line trim press can be used to cut out certain cavity areas that are not in the same plane as the trim perimeter of the finished part. The partially trimmed shot then can be guillotined and the outer perimeter can be die cut using a manual trim press.
- A multiple-cavity shot may be more efficiently processed as a whole unit and trimmed after the cavities are filled with product and sealed. Examples may be seen in food or medical packaging, such as individual jam packages or sterile single-dose tablets—both of which are packed, heat sealed, and trimmed.
- An untrimmed multiple-cavity shot offers ease of handling as a single unit for labor-intensive, secondary operations, such as post silk screening, printing, hot stamping, or radio frequency (RF) heat sealing and then final manual trimming.

Manual trim presses of varying hydraulic or mechanical designs can be seen in operation at many thermoforming plants. The ease of feeding and removing trimmed shot to and from the die determines the relative efficiency of the trim process.

## Safety

Manual punch-and-die trimming requires that the die is mounted in a power press semi-permanently and a formed cavity or cavities are manually placed in the die. When the press cuts the parts free, the trimmed parts fall through the die into chutes or a container, and the scrap skeleton is removed for recycling. This process requires safeguards as the operator's hands are often close to the die area.

## Knife-like Dies

Multi-cavity, knife-like dies can be mounted on a drawer or shuttle table, which is automatically or manually retracted and extended into the trim press with the formed shot placed in the die (see Figure 5-3). After trimming takes place, the drawer extends out of the press and the shot. The trimmed parts remain lightly attached to the shot and are manually removed as a single unit for later packing.

## Die Cutting

Two classes of manual trim presses are available for knife-like die cutting, which use either a soft anvil or hard anvil for the trimming process.

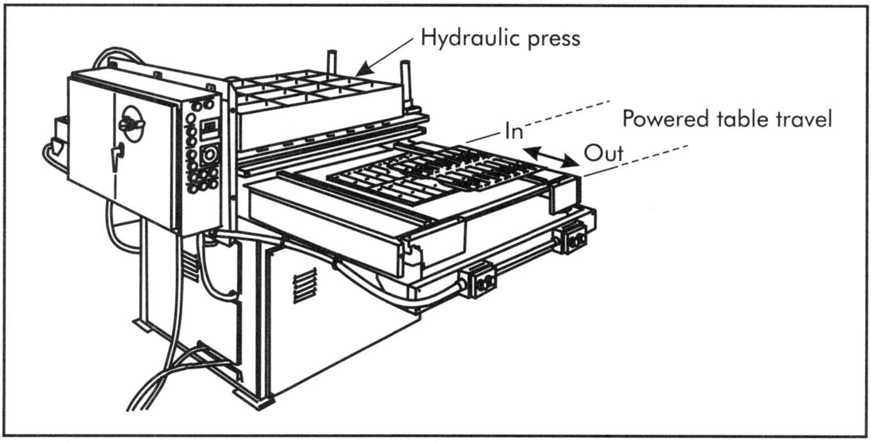


*Figure 5-3. Multi-cavity, knife-like dies can be mounted on a drawer or shuttle table (Mitchell 1998).*

## Soft Anvil Striker Plate

One type of press allows the knife-like die to cut against a soft anvil (composite rubber or plastic). When the sharp edge cuts through the plastic, it penetrates the anvil a short distance, leaving a small cut groove in the anvil (see Figure 5-4). The die can not cut continuously in the same spot on a soft anvil because the knife-scored groove will widen, causing an unacceptable trimmed

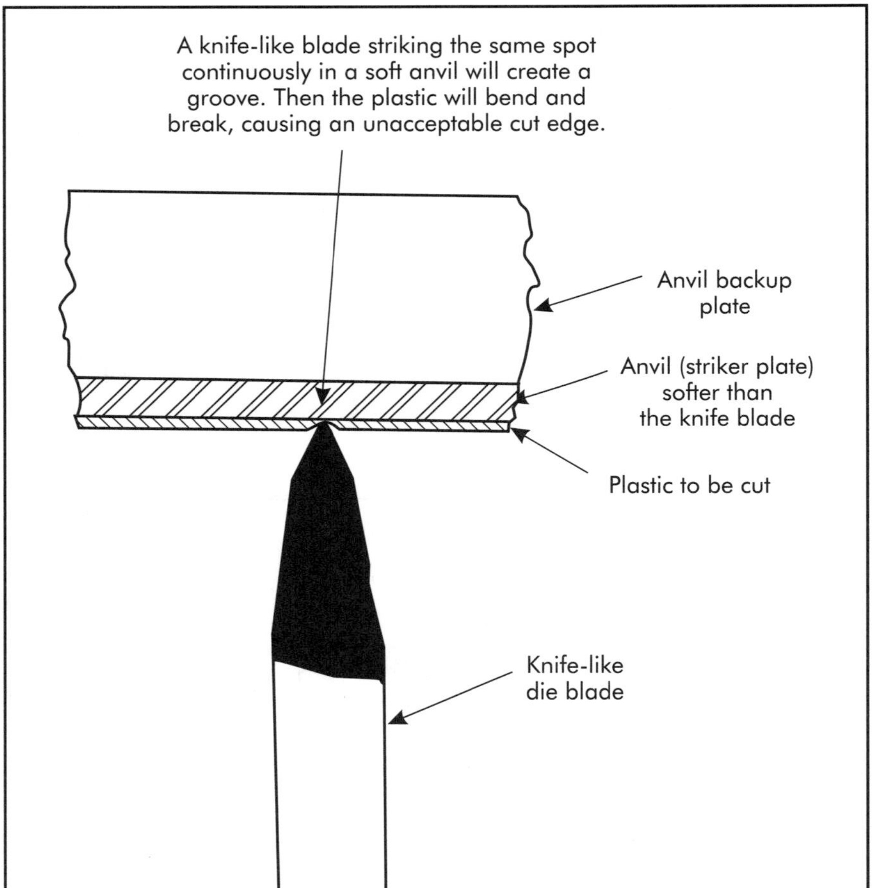


*Figure 5-4. When the sharp edge cuts through the plastic, it penetrates the anvil a short distance, leaving a small cut groove.*

part. The soft anvil or die must constantly be manually moved to avoid cutting in the same spot. The knife tends to remain sharp since it is harder than the anvil and dulls very slowly.

## Hard Anvil Striker Plate

The best trimmed edge produced by using knife-like dies results from a *kiss cut* in which the knife touches, but does not score, a metal striker plate or hard anvil. The trim press that produces this type of cut must have a defined stroke or a hydraulic-pressure control system that can be set for just the right cutting force. The finest knife-like die setup on the perfect press platen will not cut 100% of the shot until it is made ready. *Make-ready* is a system of shimming out low points on the die or reducing high points of the die and resharpening the affected blade. This is work for the most patient soul in the shop and takes time and skill, but it produces clean-cut parts and extends die life immeasurably.

## Heat Assist

When a kiss-cut trim system is available, a heat-assisted trim system may be employed since plastic loses its shear strength precipitously when heated. The striker plate is attached to a temperature-controlled hot plate, which is heated to just below the point at which heated plastic will stick to the anvil plate. The trim press is closed under low pressure until the knife-like die is bearing against the plastic, which impinges on the heated striker plate only at the knife contact line. After a short, timed heating period, pressure is increased and the die cuts cleanly through the part. It is not practical to attempt to drive heat up the thin knife-like die to achieve similar results to those obtained using a hot striker plate.

Heat-trimmed parts offer several advantages when compared to cold impact-cut parts. The press tonnage to trim a shot can be considerably reduced so that trim-die cavities can be increased in number. A heat-trimmed edge is smooth and square since the trimmed edge is not partially fractured, but is cut cleanly through.

Most of the angel-hair particulate produced by impact cutting is avoided, minimizing these unsightly contaminants. Thick, thermoformed parts are often heat trimmed to prevent internal stress buildup caused by cold cutting. Such stresses may later induce breaks and fractures in finished parts.

## Roller-die Cutting

Roller-die cutting machines provide for mounting the die knife-up on a machine bed with the movable striker plate in open position and the shot placed in the die. The striker plate is closed lightly on the die and then a large-diameter roller traverses the back of the plate, producing cutting pressure along the knife line. If the die is placed so none of its cutting edge is parallel to the axis of the moving roller, the roller arc becomes tangent to only a single knife point of contact at a time. This machine system offers an inexpensive method of producing heavy cutting pressure without using a costly, massive press. One type of inexpensive die cutter uses two rollers to produce sufficient pressure to trim a plastic sheet (see Figure 5-5).

The resultant roller-press cut edge may be of low quality because there is a force component in the direction of roller travel that may shift the shot relative to the die slightly during trimming. Some roller-die cutters move the pressure roller back and forth across the die, producing an unsatisfactory double-cut edge on the finished part.

Sometimes the roller-press dies are not made ready and a loose piece of 0.188-in. (4.78-mm) thick relatively soft polypropylene sheet is placed manually on top of the die. The sheet is removed after trimming and reused for the next cut. Since the loose polypropylene sheet is never replaced in its original position, a fresh face is presented to the die for each cut until the sheet becomes worn with use.

Computer-directed trimming of industrial parts is an extensively used technique in the heavy-gage, sheet-fed thermoforming industry. The most common computer-directed trimmer is a five-axis router, which uses a high-speed cutter to cut a narrow groove through the part wall, converting the router path into chips. Simi-

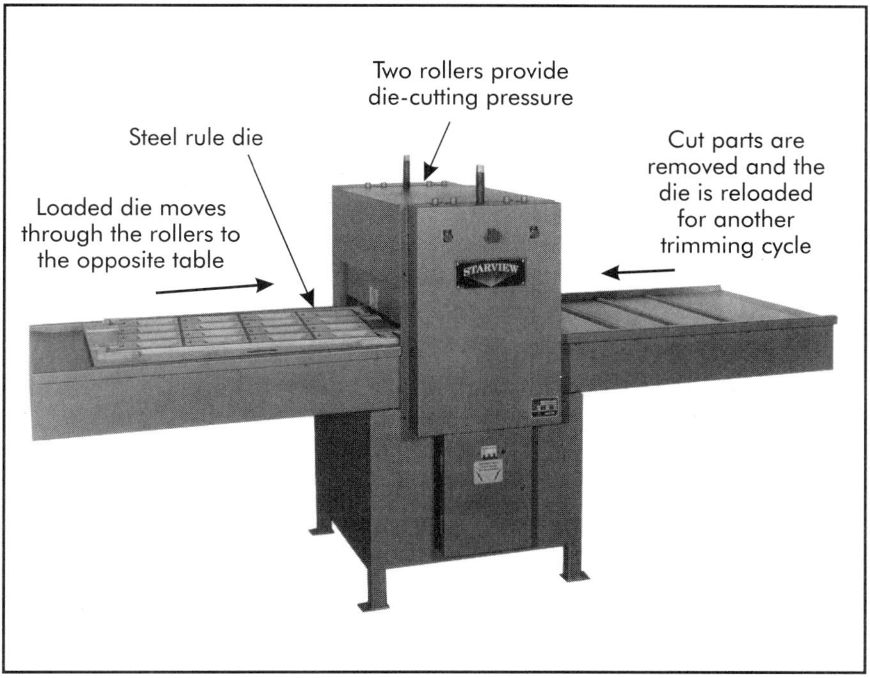


*Figure 5-5. One type of inexpensive manual die cutter uses two rollers to produce sufficient pressure to trim a plastic sheet. (Courtesy Starview Packaging Machinery, Inc.)*

lar computer-directed cutting with ultra-high pressure (50,000 psi [345 MPa]) waterjets or laser devices has been developed for trimming heavy-gage thermoformed parts. None of these systems has proven practical for cutting multi-cavity, sheet-fed, thin-gage parts because they are unable to trim parts as rapidly as a thermoformer produces them. Both systems generate particles that must be manually removed from the finished trimmed parts before they can be used.

Slitting the excess-width web sections directly off a roll-fed thermoformer is specified for specialized applications. Products that require thermoforming an endless web are trimmed to a given width by slitting. Slitting takes place as the web leaves the thermoformer and can be accomplished by razors or rotary knives. A combination of slitting and guillotine cutting the shot can produce a finished component without resorting to an outer-perimeter die.

## On-line Trim Press

On-line, knife-like, die-cutting presses built on a thermoformer frame generally contain both upper and lower moving platens (see Figure 5-6). Most thermoformed parts are formed by projecting up from the sheet line to take advantage of the upper-trim platen, with its larger clearance stroke of 4–8 in. (101.6–203.2 mm). Some parts are thermoformed projecting down to allow cementing or sealing a card to the sheet-line face. This technique completes the formed cavity and then the die cuts the composite shot. The lower platen motion has a shorter stroke of 1–4 in. (25.4–101.6 mm) than that of the upper platen. The opening clearance of both platens allows trimming parts, which combines both upward and downward projections from the trim line, and indexes through press and die.

On-line presses are often designed so that the upper-platen stroke is actuated by a toggle linkage and the lower-platen movement by direct hydraulic action or a combination of toggles and hydraulic force. This system of platen movement is helpful in piloting cavities to true position for trimming.

When a formed shot completes its index out from the mold area to the open trim press, the trimming cycle is activated. Upper toggles are fully extended to move the platen and die down to the sheet line where locator pilots in the die engage formed cavities relative to the die. After the top toggles are locked, the lower platen extends, gently nudging the sagging plastic web tightly into the die locator pilots and then building up the necessary hydraulic force to trim the shot. Indexing of the web may be quite accurate and consistent, but pilot locators in the die provide final displacement of the formed cavities to precisely trim a large multi-cavity shot.

## Conditions for Proper Die Cutting

Most trim presses and their die buildups contain too many variables for the operator to control, which affect the exact shut-height dimension for kiss cutting when using a knife-like die. Kiss cutting on the trim press can be accomplished by careful die make-ready

*Figure 5-6. On-line, knife-like, die-cutting presses built on a thermoformer frame generally contain both upper and lower moving platens. (Courtesy Brown Machine Co., LLC)*

and using minimum hydraulic pressure to trim the shot. Proper die cutting can not be achieved by use of brute force, because initial closing of the press and die may dull the sharp edge of the knife and thereafter the die will crush, not cut, through plastic. When the knife loses its keen edge, a shiny flat spot appears on its cutting edge and this flat area will crush through the plastic, leaving a trail of hairy particles (angel hair). The contaminant clings to the trimmed component due to static electricity generated on the web and it can be difficult to dislodge from the component. The end user may reject these unsightly parts so it is best to minimize angel hair by using proper trim set-up procedures.

Make-ready of a knife-like die requires the die and its buildups be made parallel to each other. Once this is done, all components should be locked into alignment. Die buildup consists of many flat plates that over years of use can become burred and battered. Often, flawed plates must be reconditioned or replaced if kiss cutting is to be achieved.

## Striker Plate Material

The skill and ability of machinery setup technicians determine the choice of a plate cutting-face material for the striker. An improperly made-ready die survives longer if the engineer chooses a soft striker plate. However, trimmed-edge quality declines as production continues. When a proper make-ready is completed, the master sheet (the sheet of paper that the make ready is attached) and die can be stored and used directly for the next production run.

Materials employed for the striker plate include:

- Aluminum—7075 alloy is one of the hardest commercial aluminum sheets and does not wear down a die, but its surface scores quickly and deep grooves result. The cut-edge quality deteriorates rapidly as the day progresses. Its estimated life is 2–8 hours.
- Stainless steel—300 series chrome-nickel soon becomes work-hardened by the die's cutting action. The die can be damaged if improperly forced into the plate, but this material

may be a good intermediate choice if the die is carefully made ready. Its estimated life is 5–30 days.

- Hardened high-carbon steel striker plates are generally several hardness points harder than the knife-like die and this plate has a hardness of 45–55 on the Rockwell-C scale. The material is flexible, but not hard enough to become brittle and crack under load. It is a very good choice for the highest-quality trimmed edge with a long life: 3–24 months when dies are made ready properly.

The thickness of a striker plate is generally between 16–12 gage (0.059–0.105 in. [1.50–2.66 mm]) as the plate must not distort readily, but should be thin enough to react to underlying make-ready shims.

The following is an example of a make-ready procedure for a knife-like die in an on-line trim press:

1. Reduce hydraulic press cutting pressure to the minimum level that will allow actuation of the lower platen.
2. Prepare a heavy Kraft paper sheet to the exact size of the platen and mark the outgoing edge as *front*. Tape the paper facing the front to the face of the striker plate, allow the press and die to close, and strike the paper, leaving a cut impression on what is now the *master sheet*. Usually 75% of the die will either cut or mark the paper. Remove the master sheet and striker plate; use a pencil to complete the cut impression of cavities that are incomplete.
3. Obtain 0.002-in. thick × 0.250-in. wide (0.05-mm × 6.35-mm) stainless make-ready shim tapes with adhesive backing from a die-maker supply house. Study the die impression on the master and apply one layer of shim tape only on the very light or penciled-in die impressions. Trim the shim tape so it never extends closer than 0.250 in. (6.35 mm) to a neighboring heavy die impression. The objective is to build up the shim pack so it never disturbs an existing cut section. Avoid installing loose shims; they may shift and disrupt the process.
4. Place the master sheet on the lower buildup in the same orientation marked "front" as the die. Install the striker plate on top of the master sheet and replace the mounting screws.

5. Tape a clean Kraft paper cut to the exact size of the striker plate on top. Mark it *No. 1 front* on the appropriate edge. Die cut the No. 1 sheet and compare the results to the master sheet under the striker plate. If the die impression still is not uniform, add one thickness of shim to the master sheet on any faint cuts, including those on the top of earlier shims. When building upon an earlier shim, cut the length shorter by 0.250 in. (6.35 mm) from each end so the shimming is feathered and not abrupt at its edges.
6. Save sheet No. 1. Then, cut sheet No. 2 and continue the process until the die impression becomes uniform. Save all the trial-cutting Kraft sheets to keep a record of progress and as a guide to avoid disturbing sections that were previously cutting. If previously cutting segments stop cutting, remove the last shims placed on the adjoining segments and start the process anew.
7. When satisfied that make-ready is complete, insert a flat sheet of the same plastic to be thermoformed and attempt to trim it at low pressure. If it appears to be a uniform impression, raise the hydraulic pressure until the die cuts through. Lock the hydraulic pressure regulator at that point and shim-up areas of the impression that may not have cut through. Judgment and experience will indicate when the die impression is uniform and when additional hydraulic pressure is needed to cut through a plastic sheet without dulling the die.

## Cause of Die Damage

Damage to the knife blade shows up as flat shiny areas or curled knife-edges, which can be felt with a fingernail. The flats on the knife can be sharpened with a stone; curled edges sometimes can be hammered out. If the damage is too severe, a spare die can be substituted until a die maker can effect repairs. The spare die should also be made ready before attempting production.

After die cutting has commenced, the knife eventually dulls due to the abrasive nature of the plastic or from impact against the die anvil. At that point, hydraulic pressure may have to be raised slightly to compensate for the worn die and allow it to continue to cut. Some plastics are notorious for causing short die life. For example, thick PVC sections cut partially through and then fracture, allowing the knife to impact the anvil. This can be recognized by listening to the hydraulic trim pressure building up. Then, a loud crack is heard as the PVC fractures and a ping indicates the die is striking the anvil. Heat trimming may reduce this problem when sufficient heating time is allotted to heat the trim line before activating hydraulic pressure.

## Off-line Trim Press

Off-line punch-and-die trim presses accept output of a thermoformer after the formed web has left the indexing chain and the web is picked up and re-indexed by the off-line trim press. Re-indexing is necessary because only one or two rows of a shot can be trimmed at one time and the shot could contain many more rows of parts. Fortunately, the elapsed travel time from the thermoformer to the trim press allows the web to shrink to its ambient state with most types of thermoformed plastics, so handbook shrinkage values may be applied to the die design.

Trim presses that direct their power stroke horizontally are described as horizontal trim presses (see Figure 5-7). They have the advantage of directing the cut and nested parts in the horizontal plane, which assists packing finished parts. The skeleton scrap web is cut off with each stroke of the die. Then, the scrap falls into a granulator or is removed by conveyor belts for recycling. Vertical off-line trim presses have some mechanical and process advantages that make them useful—often in conjunction with in-line trim presses (see Figure 5-8).

When punch-and-die trim quality starts deteriorating, a temporary repair in place can avoid removing the heavy die and stopping production. On some of these tools, the fairly soft die section can be comfortably reached from outside the press. Peening (deforming) the die edge to decrease die clearance in the defective area is something to be attempted. The much harder punch broaches the peened area of the die cavity when the press is closed. It shaves excessive metal and maintains tighter clearance.

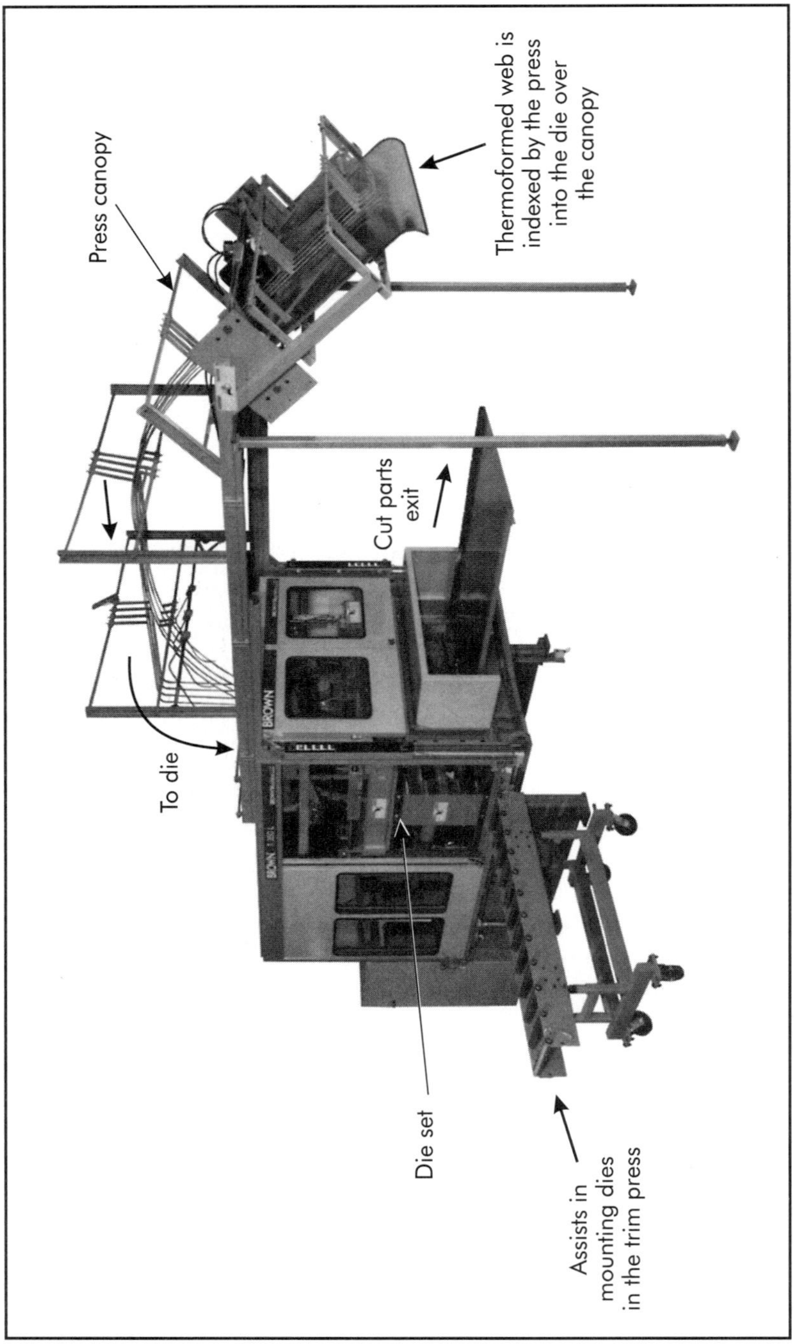


*Figure 5-7. Trim presses that direct their power stroke horizontally are described as horizontal trim presses. (Courtesy Brown Machine Co., LLC)*

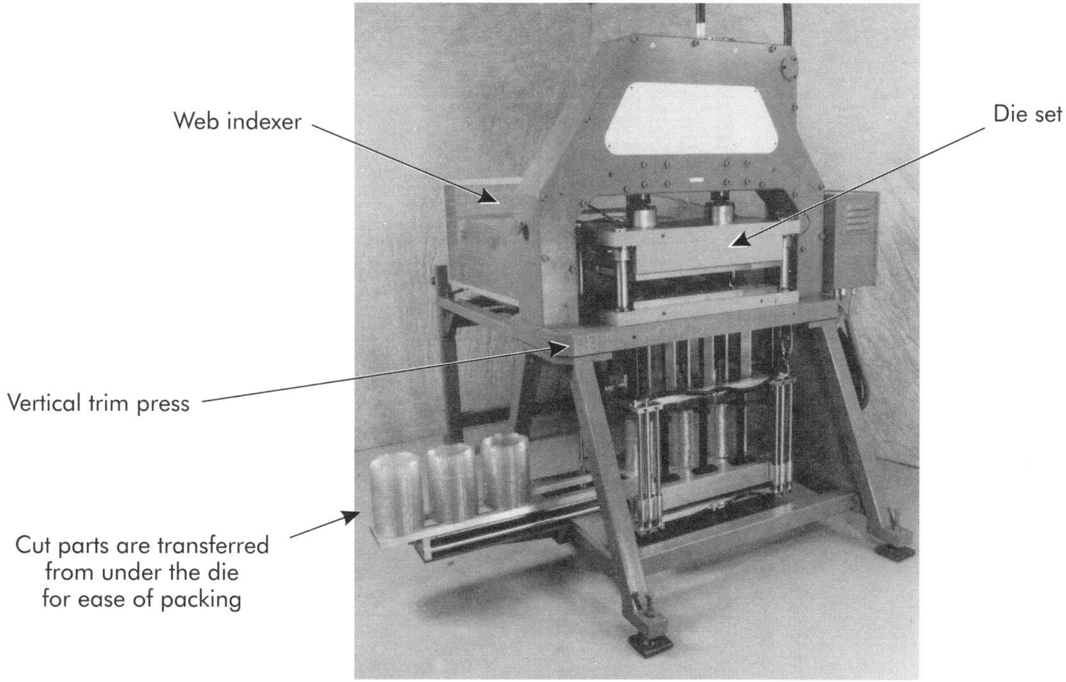


*Figure 5-8. Vertical trim presses have some mechanical and process advantages that make them useful—often in conjunction with in-line trim presses. (Courtesy E.D. Segen & Co., LLC)*

Planning at the mold design level is necessary to provide sufficient trim-press indexing and guidance clearance into the punch and die. Pentagon-shaped pusher lugs are mounted so that an indexing pusher can move the section of the shot to be trimmed into the die (see Figure 5-9). Pilots built into the punch assist in the final registration of formed parts relative to the die. Press indexing accuracy should be a consistent ±0.032 in. (±0.81 mm) for pilots to centralize the cavities. The mold design should incorporate space for index guide bars that direct the moving web into the die and prevent side-to-side displacement of the shot.

The die designer often is responsible for auxiliary devices needed for a successful horizontal trim press operation. Such devices may include:

- A web-indexing cam, usually mounted to the trim press crankshaft, provides the timing sequence for a specific index stroke on some press models. Accelerating and decelerating the web-indexing stroke provides a smooth start and stop of the web. The cam may be of a cycloid shape or have a similar curve to produce this motion.
- Knockouts are often necessary to push the trimmed part past the die lands (area of the die cavity that has minimum die clearance). This facilitates stacking parts as they pass through the die exit. Most trim presses are designed with an integral knockout platform to actuate the die knockout components at the proper time.
- Product chutes are needed to keep products in neat rows for easy packing once they clear the die. Chutes are often made of polished stainless steel to reduce friction as trimmed parts are pushed along out of the die. The stainless material is easy to keep clean, very strong, and resists corrosion.
- Spare punch and die modules may be needed to continue production in case of serious tool damage. Punches and dies go through a long manufacturing cycle including machining, heat treating, and grinding. Procurement of special tool steels to manufacture emergency spares can be expensive and time consuming. When the original punches and dies are ordered, additional spares can be run through regular die-shop production at the same time, reducing cost and delivery time.

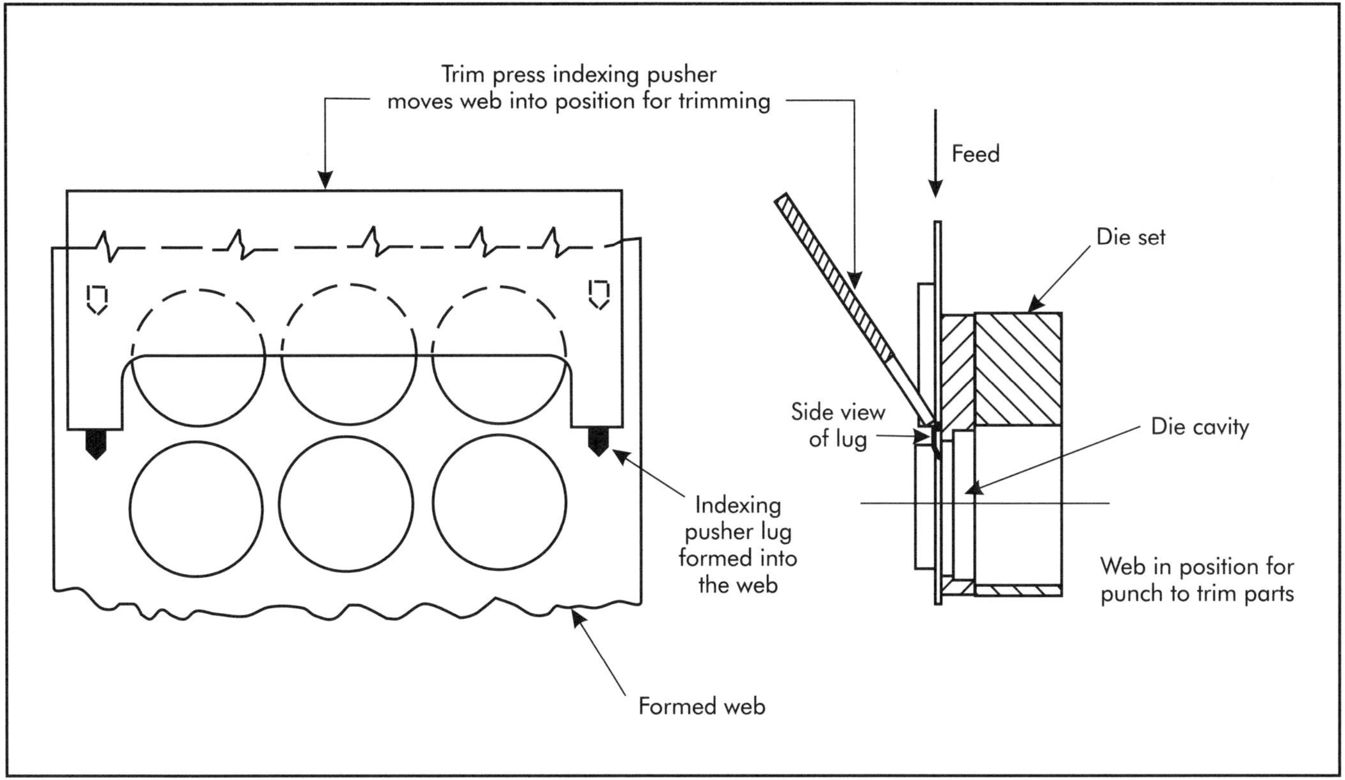


*Figure 5-9. Pentagon-shaped pusher lugs are mounted so that an indexing pusher can move the section of the shot to be trimmed into the die.*

Die makers often produce spares for their own purposes when building the die to avoid problems that may later occur during grinding or heat treatment. The user can not rely on the die shop's limited supply being made available because these extra spares may be used up to complete the die.

Web guides and index pushers can be designed and built during the design stage to match the needs of the specific parts being trimmed. Most trim presses come equipped with a general set of movable sheet guides, which can be adjusted for each die. Setting up these adjustable components is cumbersome and they often loosen and move during production. Permanently fixed components are preferred at the production level.

## Scrap

High-speed trimming of plastic produces occasional scrap components. These components, as well as the skeleton scrap web, must be prepared for recycling. On-line thermoformers may have scrap winders that wind the scrap web on a mandrel, which is removed manually on a regular schedule. Alternately, guillotine shears mounted at the end of on-line thermoformers may cut the skeletons up into individual shots, which are collected manually or by conveyor system for granulating.

Off-line horizontal-trim-press dies have built-in web cutoff blades that trim the scrap skeleton with each press stroke, after which it drops down by gravity. Many off-line trim presses are built to accept a granulator below the die. Scrap falls only a short distance and is immediately ground into small particles. Scrapped or damaged thermoformed components often are too bulky or packed too densely to pass into the throat of an in-line granulator and are handled offsite by larger and more powerful granulating equipment. When feeding a continuous web into a granulator directly from an on-line thermoformer, provisions should be made to restrain web movement to prevent the web from being dragged into the grinder and upsetting index accuracy.

## Trim Press Selection

The different trim systems described in this chapter serve various production situations and often more than one trim system can be substituted for another. Usually, the choice of trimming comes down to an economic decision based on initial die cost, labor burden, skill level of technicians, and available trim machinery. Sometimes, trimmed product specifications are so exact that only one trim process can be employed to achieve a given result and economics must take a back seat.

Engineering should follow up on the results of selecting a specific trim system to guide thinking for future projects and judge the efficacy of each process. Production employees may be able to offer advice that may reduce cost, increase production, or they may advise that the project was a nightmare producing an inordinate amount of scrap. Usually, the truth lies between the extremes of machinery technology and the skill level of operators and their supervisors. The engineer is charged with getting the job done and keeping a balance between tool technology, existing trim equipment, and the abilities of tooling users.

## Reference

Mitchell, Philip, ed. 1998. *Tool and Manufacturing Engineers Handbook*, Fourth Edition, Volume 9: *Material and Part Handling in Manufacturing*. Dearborn, MI: Society of Manufacturing Engineers.

# 6

# Mold Design

## Considerations

The mold design process requires defining the mold objective, choosing the proper plastic resin, finding the critical wall thickness, determining if the fit is too tight or loose, testing for thermoformed part clarity, deciding if hinges are used, how the mold will be engraved, how many cavities are needed per shot, and the height of the thermoformed product.

### Defining the Mold Objectives

The design of a thermoforming mold that produces the best product for the customer at the least cost depends on the designer obtaining as much information as possible at the project start. Models, drawings, and prototypes provide a good beginning if available, but they should not end the search for facts. A customer may ask the mold designer to develop a thermoformed package to fit within a given space. He or she may supply an assortment of products for placing into the package and ask the designer to develop a package to fit within a given space. This provides the opportunity to develop the part with proper thermoforming- and trim-design parameters, with the designer working out the details with the customer. Direct conversation with the mold purchaser may unearth facts that bear directly on the final design, but were not covered by drawings or a salesperson's memory of initial sales conferences.

Part details should be reviewed with the mold purchaser to ascertain which elements are most crucial and which have secondary importance. Mold information is only part of what should be discussed with the buyer before mold design work can begin. It is important to investigate every area of the project that may or may not be directly the responsibility of the mold designer. The designer's initial blunt questioning will be forgiven once the tool begins production on time, within budget, and with no bitter aftertaste.

## Choosing the Proper Plastic Resin

Price should not be the deciding factor in selecting a thermoforming plastic. The ideal resin with proper strength, clarity, or cushioning properties may first appear too expensive on a pound basis and, therefore, may not be considered an option. An expensive resin may meet price criteria if its greater strength allows reduced wall thickness or modification of its structure to stiffen the thermoformed part.

Chemical compatibility with the product being packaged should be investigated. Most plastic resins have varying degrees of porosity to atmospheric gases that eventually permeate through package walls and may degrade a food product.

Chemical additives mixed with plastic resins to enhance the physical properties of the resin may migrate out of the sheet and damage package contents. Government regulations may direct use of specific resins for foods and medical products due to possible chemical contamination of package contents.

High temperature can cause thermoformed packages to distort. For example, medical products may require steam autoclaving during further processing. Solid foods may be hot-filled into a thermoformed package; the plastic resin chosen should have a sufficiently high heat-distortion point to prevent deformation during filling. The initial resin specified determines the mold and die shrinkage coefficient. A late change in material can cause a variation in finished package size that requires new mold cavities and dies.

## Critical Wall Thickness

Thermoform processing, unlike injection molding, cannot assure uniform wall thickness throughout a part. However, if minimum wall thickness is specified in specific areas of the part, the designer can choose the mold and thermoforming technique needed to achieve this result. Thoughtful mold design is a major priority. It should have the goal of creating approximate uniform wall thickness throughout a thermoformed product.

Product design of the formed part should be in harmony with the process used to assist flow of plastic uniformly over the mold surface. The draft angle of the walls and all radii should be sufficiently generous to help produce the minimum specified wall thickness.

Heated plastics move more freely on a hot mold than on a cooler one, but the maximum production rate often is based on using a low-temperature mold. When the mold temperature is too high for an effective plastic flow to achieve minimum wall thickness, the required extended cooling time can result in uneconomical production.

The starting sheet thickness for a minimum thermoformed wall must be determined in advance of production so that plastic sheet rolls can be extruded in time to maintain the production schedule. Prototype mold testing is imperative to determine the optimum plastic sheet thickness used in production.

## Is the Fit Too Tight or Too Loose?

Many thermoformed parts are designed to fit or lock on existing units. Lids for drink cups and locking salad trays are typical thermoformed parts subject to judgment calls on tightness-of-fit. Since tightness is subjective, correct fit can create demands for modifying completed molds before the true cause of deviation is investigated. Often the difference may be due to variances in mating parts. Matching components, such as foam coffee cups, may vary in diameter. Cardboard assembly elements can shrink or expand depending on relative humidity. Consequently, thermoformed or injection molded units to be mated may not be to specified size.

It is important that the designer initially be given a number of mating parts so a random sample of prototype thermoformed components can be tested. Approved thermoformed package samples should be marked, dated, and initialed by the customer, and stored by the designer. Any future disagreements can be negotiated using initialed samples as the basis for discussion.

When designing mating fits for thermoformed parts, it is always best that surfaces in contact be formed directly against the mold (see Figure 6-1). When this procedure is followed, wall thickness will not be a factor in dimensioning the final fit. Wall thickness in the fit area can fluctuate due to variations in sheet-heating pattern, shifts in mold temperature, or a change in starting thickness of the plastic web. A modification of a mold cavity as small as 0.001–0.002 in. (0.03–0.05 mm) on a coffee cup lid can cause the dreaded words *too loose* or *too tight* to be uttered.

## Part Clarity

Part specifications may request water-glass-clear thermoformed parts to allow potential purchasers to view merchandise in the packages, but thermoforming can cause loss of sheet opacity. When

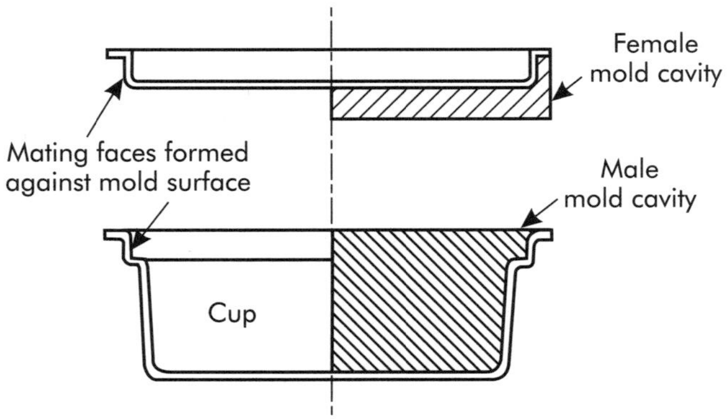


*Figure 6-1. Consistent fits of mating parts surfaces are achieved by forming directly on a mold face.*

clear plastic sheet is specified, the designer should attempt to minimize any reduction in clarity caused by the mold. A designer should pay attention to mold surface finish, placement of vent holes, and how and where to stiffen package structure without distorting the view through the package.

Thermoformed plastic needs vent holes or slots to quickly evacuate air trapped between a heated thermoforming sheet and the mold surface. Drilled vent holes may be reproduced on the finished, formed plastic surface, but often are objectionable to the customer (see Figure 6-2). Package-stiffener ribs and decorating

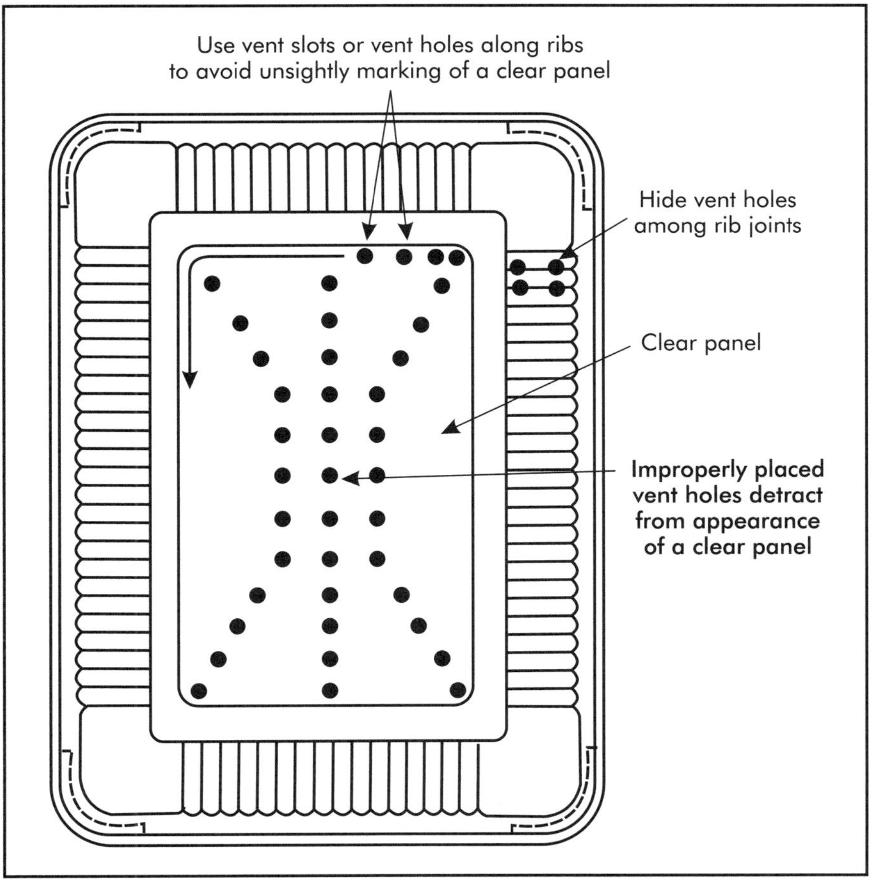


*Figure 6-2. Careful placement of vent holes avoids marking a clear panel. (Courtesy G.N. Plastic Co., Ltd.)*

trim can hide these holes, or vent slots can be located at changes of mold section. When the only vent holes are on the perimeter of a thermoformed panel, the surface finish of the mold determines efficient evacuation of entrapped air. A rough surface finish directs more airflow through passages in the hills and valleys of the surface to the vent holes, but decreases package clarity. The designer's dilemma is to either move the holes into the visible area or provide a rough surface finish that reduces clarity.

Large panels of unstiffened, thermoformed plastic in a flat plane become limp and distorted, no matter how much additional plastic is added to the starting sheet thickness. Redesigning a large panel with a series of layers stiffens the structure yet still allows some package visibility without increasing wall thickness.

Trapping air between the mold and the plastic is a thermoforming technique sometimes used to achieve clarity on a flat thermoformed panel (see Figure 6-3). The designer creates a mold without vent holes on the panel mold surface so that the hot plastic does not contact the mold surface as it floats on a layer of entrapped air. Since the heated panel never comes in direct contact with the

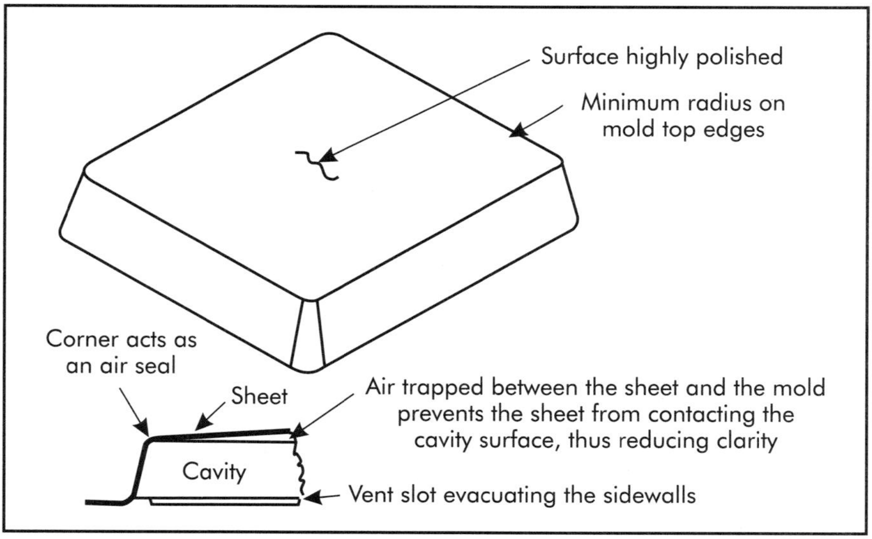


*Figure 6-3. Thermoforming a clear panel surface by trapping air between the sheet and cavity.*

mold, it cools more slowly and remains hot after it is stripped from the mold. As it indexes out of the mold, the heated panel may distort as it cools in the air.

## Hinges

A thermoformed package consisting of two halves with a formed hinge in between may be reused to store such products as fasteners and fresh spices. The buyer may wish a soft hinge for opening or closing instead of a stiffer hinge to prevent the package from requiring both hands to keep it from spilling its contents. A hinge can have a softer action if double V-shaped (W) grooves are specified rather than a formed radius groove (see Figure 6-4). When heated plastic is thermoformed down into a W-shaped hinge groove, wall thickness tends to thin out, decreasing the tendency of the package to remain open, whereas a higher-stiffness (radius or "V") hinge acts as an opening force. Additionally, a W-shaped hinge prevents the rear locking area of the package from springing open.

The main disadvantages of a double-V hinge are that it takes up additional room on the mold (uses more plastic) and may not have the same aesthetic clean look when the package is closed. The buyer should be made aware of hinge options before the design is frozen; later mold alteration may be difficult.

## Engraving

Engraving molds with lettering or graphic designs requires both buyer and seller to understand how the final product will appear. The sharpest quality of engraving on a plastic component is accomplished when the engraved face is thermoformed against the mold. When the engraved face must be viewed through a thickness of opaque plastic, its sharply defined features may become rounded or unclear. If the plastic is of a clear material and the engraved face is plainly seen looking through the wall thickness, the results can be acceptable.

Engraved portions of a mold are cut to read as they appear or are engraved as mirror images of written text (see Figure 6-5).

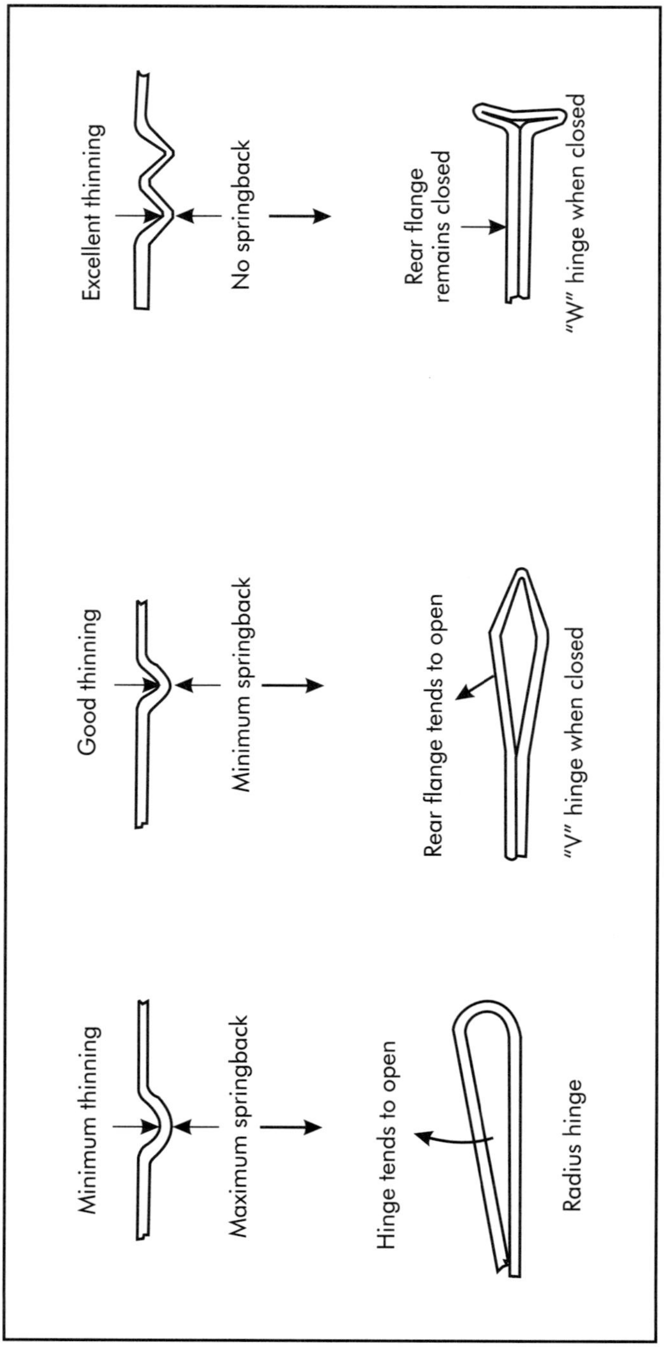


*Figure 6-4. Radius, "V," and "W" package hinges.*

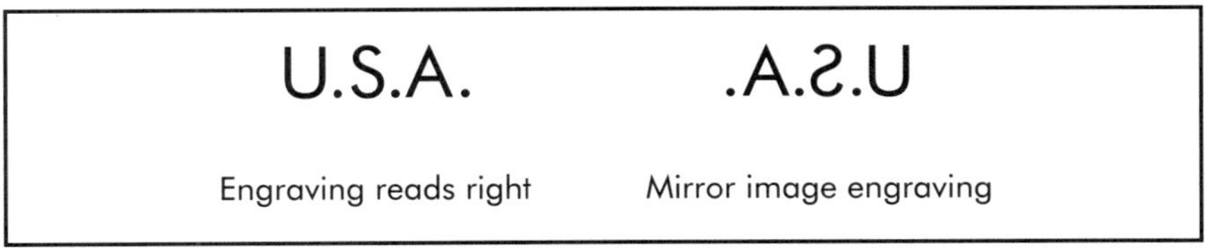


*Figure 6-5. Engraving can read right or as a mirror image.*

The engraving can be read in the cavity directly, but will appear as a mirror image when viewed from the mold side of a thermoformed part. Engraving is legible when viewed through a clear wall thickness face; the opposite is true of mirror image engraving. Raised, engraved-cavity lettering forms female inscriptions in the plastic facing the mold surface; sunken engraving appears raised on the plastic face.

Engraving details should be carefully discussed and final agreement should be spelled out in written form. It is customary for the buyer to supply the mold designer with copy, logos, typeface, and graphic designs. Spelling and punctuation of the copy should be checked before delivery.

## Number of Cavities per Shot

The mold customer specifies the type of thermoforming machine to be used for production and the number of cavities per shot expected. Machinery manufacturers or the customer can provide mold-mounting dimensions, maximum mold size, and other data required to fit the mold to the machine. Arrangement of cavities in the shot is part of the layout procedure, which is discussed later.

It is important to make every attempt to satisfy buyer expectations regarding the number of cavities required, since this quantity is the basis of the mold buyer's quote to the end user. Many mold buyers push the limits of credibility about the number of cavities to be placed in a given mold area. However, if the designer prepares a study layout indicating practical cavity placement, the buyer should carefully consider this information before proceeding with mold design to avoid future thermoforming problems.

## Height of Product

Clearance between the thermoformed cavity and the maximum opening cited by equipment specifications could cause a problem to which a joint solution should be reached with the customer. Machinery manufacturers specify the maximum open and shut heights of their presses. Shut height is equal to adding together the upper- and lower-mold assembly height dimensions. When the press is open, the space to the sheet line equals the maximum clearance for the formed web to pass. If a maximum-height shot of formed parts is rigid enough not to sag along the sheet line, the maximum press-clearance dimension can be relied upon.

An incoming sheet heated to thermoforming temperature expands in width and sags down. The degree of sag is based on the type of resin specified and the width of sheet. If the hot sheet sags and touches a male mold or the plug assists mounted on the lower platen, this contact may become an obstacle to proper thermoforming. The heated sheet can become chilled by contact with the tool and unsightly drag marks may show up on finished parts. This condition of the hot web touching mold components located on the extended lower-platen mold occurs with in-line thermoforming and die-cutting machines because the top trim platen usually has a longer stroke than the lower platen. Therefore, parts are thermoformed while projecting upwards to clear the upper-platen tooling. Once the decision to form parts upwards is made, the plugs or male molds must be mounted on the lower platen.

Some thermoforming machines contain auxiliary sag bands—narrow, thin, endless steel bands similar to a bandsaw that move with the hot web through the process and then cool as they return to the oven (see Figure 6-6). Sag bands reduce web drooping by narrowing the unsupported length of the web. Clearance for band passage through the mold area should be provided in the mold layout. Machinery not equipped with sag bands may reduce sag by shortening the web width or by using a resin that has greater hot strength or modulus and, therefore, less deflection.

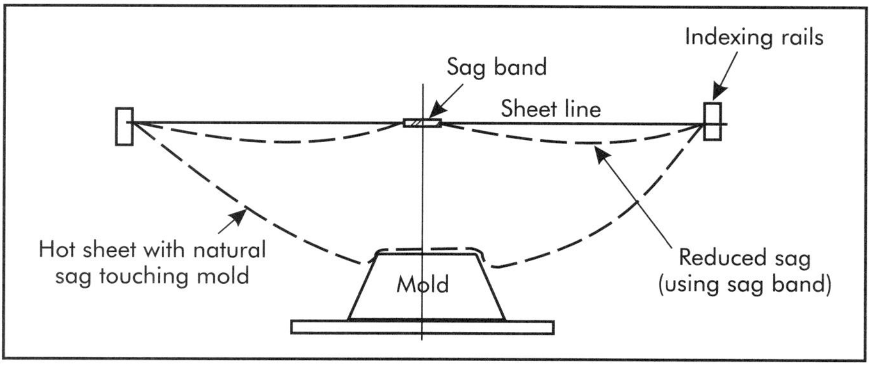


*Figure 6-6. Heated sheet sagging onto lower tooling can be reduced by using a sag band.*

## Mold Cavity Materials

Unlike injection, rotational, blow, or foam-bead molds, all of which require two mold halves, thermoforming—with the exception of twin sheet and foam sheet—needs only one mold half; the sheet thickness face becomes the back surface. Injection molds are made of heat-treated steel. They may require complex heavy machining and many added costly mechanical components to function properly. Simple thermoforming molds can be made of aluminum-filled epoxies, or from stock or cast aluminum. These are relatively easy to machine and can be fabricated quickly and inexpensively. Economic factors helped promote initial interest in thermoforming. Unfortunately, labeling it a "poor man's injection molding process" proved to be a burden for the thermoforming industry as it matured.

The first multi-cavity thermoforming molds were fabricated of castings of liquid epoxy resins or aluminum because these techniques were in common use in the 1950s before computer-directed machine tools were available. Both casting techniques require a usable model or carved wooden pattern as a starting point for the procedure. The wooden pattern maker's tradition of adapting to new ideas provided excellent guidance for early thermoforming mold builders.

## Aluminum-filled Epoxy Cavities

Aluminum-filled epoxy cavities are often selected for in-house mold making because they are easy to fabricate or can be bought from a vendor and then mounted to a mold base. Aluminum-filled epoxy is composed of two liquid components of plastic that are mixed just before a mold is poured. The mold is then air cured until it becomes rigid in 12–24 hours. Many epoxy formulas require an additional oven cure to assure the molds become sufficiently resistant to high thermoforming heat. Fully cured epoxy cavities machine easily, maintain a suitable surface finish, and have little shrinkage. However, aluminum-filled epoxy possesses poor heat transfer characteristics when compared to solid aluminum, so that its maximum cyclical thermoforming speed varies between two and five shots per minute, depending on mold thickness. This slow speed was not an impediment to early model thermoforming machines because they were not capable of operating at higher rates of production.

### *Model or Pattern for Casting*

Before a cast mold can be fabricated, a model or pattern must be suitably prepared by a pattern maker. The finished pattern for an epoxy casting is enlarged to compensate only for plastic sheet shrinkage. Aluminum castings have shrinkage of 0.013 in./in. (0.013 mm/mm). The specified plastic shrinkage must be added. For example, total shrinkage for high-impact polystyrene (HIPS) on an aluminum casting would be:

0.013 in./in. (0.013 mm/mm) aluminum + 0.005 in./in. (0.005 mm/mm) plastic = 0.018 in./in. (0.018 mm/mm)

The model or pattern should reflect this increase in size. A pattern maker adds suitable radii, draft angles, and fillets to complete the pattern.

A multi-cavity mold using aluminum-filled epoxy castings can be manufactured using only one single prepared model or pattern. This is done by thermoforming many thin plastic shots using the pattern as a mold. These throwaway, formed shots are used as individual patterns for the final multi-cavity mold. If this technique is employed, it is best to specify double sheet shrinkage to calculate pattern dimensions because plastic shots used as a pattern shrink in addition to the finished thermoformed product.

It is important that the model or pattern have a smooth finish without any undercut faces, which cause problems in stripping finished throwaway shots. Wooden patterns intended for direct thermofoming of shots to be used for an epoxy casting do not receive paint or varnish coatings because such coatings soften and adhere to throwaway thermoformed shots. Aluminum casting patterns are finished and coated in the traditional manner to prevent moisture being absorbed from foundry sand. The best wooden patterns display minimum wood grain and are made of poplar, maple, fabricated plastic, or aluminum stock.

Epoxy tends to cement itself to whatever it touches, so release agents are applied liberally to throwaway plastic shots to facilitate removal of finished epoxy cavities. When the two components of epoxy are mixed, exothermic heat is generated which, if the temperature is allowed to become too high, distorts thermoformed throwaway shots and deforms the finished epoxy molds. A low exotherm specification is ideal when choosing an epoxy formulation, unless other means of extracting heat can be employed. When fastening epoxy cavities to a mold plate, it is prudent to use a commercial, threaded-metal insert to accept the screw because epoxy threads strip easily.

Producing epoxy castings requires an in-house group that is experienced in handling epoxy and patterns. Epoxy cavities are used for quick delivery and low-cost tooling. The thermoforming rates of production with epoxy castings are less than one-quarter of that for similar metal molds due to their lower heat conductivity.

## Aluminum-cast Cavities

Aluminum-cast castings are best poured in fine sand at foundries that specialize in thermoform castings. The pattern supplied to the foundry should be mounted on a match plate of the size specified by the foundry. Today, not many foundries handle unmounted patterns as they strive for mechanization to simplify their jobs. Fine sand is recommended for mold cavities to avoid porosity, which may appear as surface blemishes in finished aluminum castings. Many molds are cross-drilled for water cooling and casting porosity; this may cause water leakage during use and disrupt the thermoforming process.

The specialized mold foundry tends to provide castings with relatively clean surfaces, thereby reducing mold-finishing labor. The cost of hand finishing the surface of a mold can easily be greater than the price of the castings, so high-quality castings are a worthwhile investment. There are casting techniques that use plaster or lost-wax-process investment materials other than sand to produce excellent surface finishes, but these castings often are porous just below the subsurface. Trapped gases retained in cast metal cause this porosity. Other casting mediums less permeable than sand tend to reduce the escape of gases. When these types of cast cavities are polished, porous blemishes may be exposed and ruin the cavity.

To produce aluminum castings requires a high-quality mold foundry and a pattern maker who is familiar with molds. An in-house tool group or an outside vendor can assemble finished cavities into a completed tool.

## Machined Cavities

Machined mold cavities provide the most accurate method of reproducing multi-cavity molds directly. High labor input made early aluminum cavities expensive because they were fabricated by manual tools and required complex machining setups. Modern computer-assisted machine tools allow an original computer design program to be modified for tool shop use in directly manufacturing a cavity. Once the machining program for the cavity is available, one prototype or 100 production cavities can be produced with great accuracy and at minimum cost.

A summary of the advantages and disadvantages of various mold cavities used for thermoforming is given in Table 6-1. Cavities are

Table 6-1. Comparison of thermoforming cavity materials

| Cavity Type | Epoxy Casting | Aluminum Casting | Machined Aluminum |
|---|---|---|---|
| Accuracy | Acceptable | Acceptable | Highest |
| Heat transferability | Lowest | Medium | Highest |
| Overall cost | Lowest | Medium | Highest |
| Delivery time | Fastest | Medium | Slowest |
| Availability of vendors | Lowest | Medium | Highest |
| Cost of surface finishing | Lowest | Highest | Medium |

the heart of the mold, but the completed tool requires many other components, such as a water-cooled mold base, pressure box, mold assists, or plugs. The question of how deeply a thermoforming mold buyer wishes to be involved in mold manufacture requires an analysis of his or her overall business plan. In-house mold shops can be an asset for every thermoforming operation.

Machined cavities require expensive computer-assisted machine tools, programmers, tool designers, and toolmakers to build complete tools in-house. Outside mold makers often cooperate with an in-house tool group to supply portions of a mold as a joint project as long as they have a fair opportunity to bid on other molds.

## Mold Shops

A Grade 1 in-house mold shop complements a thermoforming operation where all molds are purchased from outside vendors. It requires a minimal tool facility. This shop should be capable of repairing minor mold problems, modifying plugs and mold assists, and completely assembling and disassembling molds. The shop equipment would include simple hand tools and some common machine shop equipment operated by a skilled mechanic or machinist.

A Grade 2 in-house mold shop mounts purchased cavities to a mold plate. It should be able to produce epoxy molds, auxiliary plugs, and assists. This operation needs a knowledgeable shop supervisor, machinists, and manual machine tools.

A Grade 3 in-house mold shop builds complete molds. It requires mold designers, tooling programmers, and toolmakers, plus a complete computer-assisted machine shop. The financial investment is similar to that required to open a commercial mold shop. An experienced manager is key to its success.

## Mold Cavity Design Criteria

The thermoformed components, manufacturing, and material choices allow engineering to proceed with selection of optimum criteria for mold cavity design. Roll-fed thermoforming uses two thermoforming methods requiring either male or female mold cavities—except when forming twin or foam sheets, which require two-piece matched molds. Sheet-fed molds, which produce heavy-gage products and have a cycle timed in minutes rather than seconds, may incorporate techniques not used for roll-fed thermoforming. The thermoformed product specifications influence the choice between the less expensive male cavity and the more costly female cavity.

### Male Mold Cavities

The cost of machining, venting, finishing, engraving and mounting a multi-cavity male mold is approximately 50% less than that for a comparable female mold. The male cavity is completely open on all its faces for any manufacturing operations that may be required. This remains true for both cast and machined molds (see Figure 6-7). Completed male cavities can be fastened to a mold plate or to a water-cooled mold base. If less than 2 in. (50.8 mm) high, they can be efficiently temperature controlled. Female molds require mold makers to work in a hole, with easy access mainly from the open face. In addition, the outside mold perimeter must be machined to size so all cavities nest together when assembled.

Venting a male cavity often does not require drilling many vacuum holes; it may need only slot venting at its base. A very

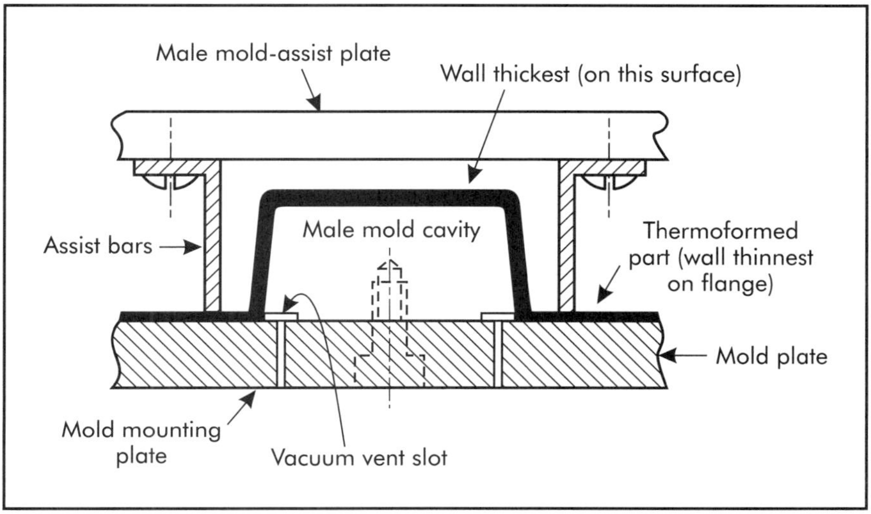


*Figure 6-7. Male mold cavity indicating the natural distribution of plastic wall thickness (Mitchell 1998).*

important advantage of the male cavity is its ability to be easily shifted to accommodate any die registration inaccuracy that appears in production. Female molds require machining of the abutting cavity walls to move them closer together, or enlarging the complete mold if cavities are to be separated (see Figure 6-8).

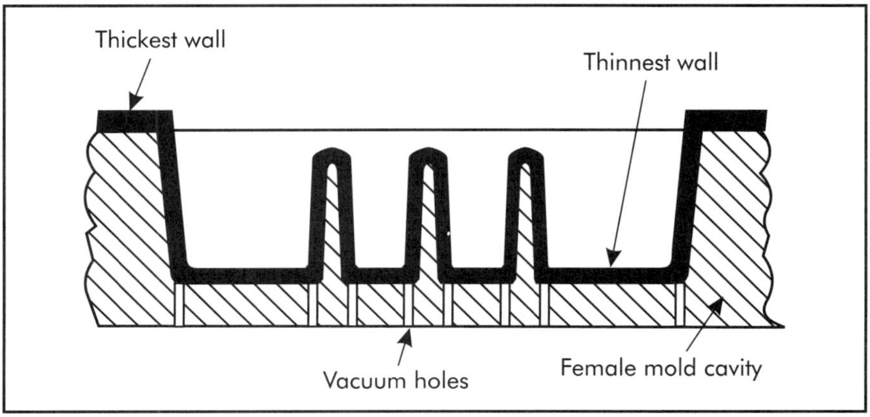


*Figure 6-8. Female mold cavity indicating the natural distribution of plastic wall thickness (Mitchell 1998).*

## Female Mold Cavities

Female molds are often used where the thermoformed parts have a depth to minimum mold opening dimension ratio greater than one—for example, soft drink cups. When an undercut is part of the product design, a female mold allows the plastic undercut portion to shrink away from the cavity—not contract and tighten its grip as on the male mold—permitting easier part removal. Layout of female molds permits closer center-to-center mounting of cavities than on a male configuration, increasing the number of parts per shot. Engraving and texturing of female cavities produces sharp, crisp reproduction of the outside face of thermoformed parts. With a male cavity, the engravings lose some detail due to mold wall thickness.

## Wall Thickness Distribution in Cavities

Natural distribution of a thermoformed sheet provides thicker walls in portions of thermoformed parts that are 180° opposite on male and female cavities. A male part is thicker at its top face and a female cavity is thinnest at its greatest depth unless auxiliary sheet-distribution devices are employed to uniformly distribute wall thickness. When hot plastic sheet impinges on the first cavity surfaces it touches, the plastic walls tend to chill and remain thicker. It will not stretch and slip as readily as non-contacting material, which is hotter and tends to thin out when stretched.

Male molds without auxiliary assists naturally tend to produce parts that are thicker at the cavity top and thin out as the sidewalls approach the flange areas. Wall thickness can be more uniformly distributed by using assists mounted in the pressure box or to the platen opposite the mold (see Figure 6-7). These assists contact the web as the top surface of the male cavity extends to the sheet line. They redistribute the plastic wall as if it continues to travel up through the opening created by the assist. Assists discourage formation of webs, which appear wherever there is excess hot plastic, especially in corners of rectangular male molds.

Female molds naturally form parts with thin bottoms at the maximum depth of the cavity because thicker top flange areas are restrained and sidewalls thin out toward the bottom of the parts (see Figure 6-8). Redistribution of female-cavity wall thickness requires use of a plug mounted on the platen opposite the female cavity. Chapter 8 describes plug geometry and the most efficient working temperature for a plug.

## Maintaining Correct Mold Temperature

Heat transferred from the plastic web to the mold must be extracted from every shot to maintain the mold at prime temperature for thermoforming. The optimum mold temperature for each plastic resin varies widely from 60° F (16° C) for commodity plastics, such as high-impact polystyrene (HIPS) and polyvinyl chloride (PVC), to 300° F (149° C) for polysulfones. Various heat control systems can satisfy this wide variation in optimum temperature requirements for molds.

### *Air Cooling of Molds*

Air-cooling of the sheet was first used when low production rates of one to two cycles per minute were typical of early-model, roll-fed thermoformers. These early forming machines were equipped with fans and optional fogging-type water-spray guns, which worked efficiently at the beginning of the day, but as the mold gradually heated up, production slowed proportionately.

### *Water Cooling of Molds*

As molds became more sophisticated and were designed with internal water-cooling passages, city water was circulated through them. City water in most areas is expensive and its temperature varies—cool in the winter and warm in the summer. The mold heat-transfer rate is dependent on the season.

Water temperature controllers for molds maintain a given water temperature by heating or cooling fluid pumped to the mold (see Figure 6-9). The water temperature is raised by electrical heating elements and cooled as it circulates through a heat exchanger. The heat exchanger uses low-temperature cooling fluids controlled by a temperature-modulating valve that monitors the mold temperature output. This cooling liquid consists of a glycol and water mixture circulated by refrigerating chillers or flowing from rooftop evaporative coolers. Water temperature controllers have a temperature range of approximately 60–250° F (16–121° C); oil temperature controllers can operate to a maximum of 600° F (316° C).

If there is a leak in a water temperature control system at a temperature above 212° F (100° C), the pressurized, heated water

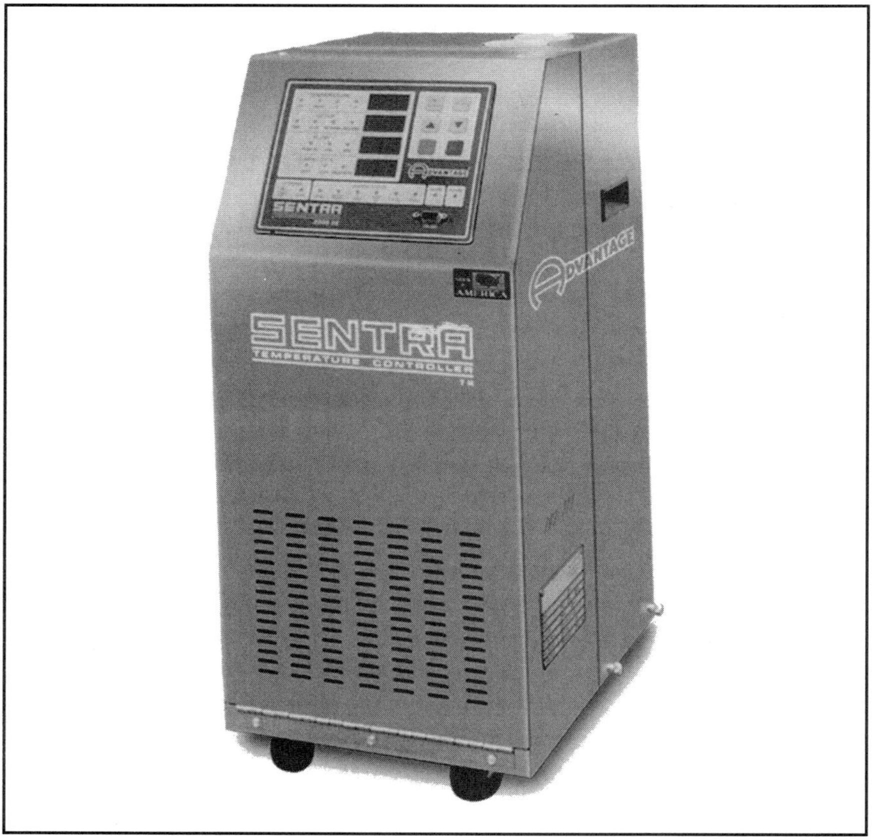


*Figure 6-9. Mold water temperature controller. (Courtesy Advantage Engineering, Inc.)*

flashes into steam that can injure nearby employees. Hot oil leaks remain liquid, but may cling to the skin and cause serious burns to anyone unfortunate to be in the vicinity. Therefore, many mold designers who require temperatures above 250° F (121° C) prefer to embed electrical heating elements into molds or plugs, with temperature sensed by thermocouples connected to temperature controllers.

### *Electrically Heated Molds*

Whenever specifying heating elements for molds, the lowest voltage available should be used to reduce the danger of accidental electric shock. The standard voltages commercially in use for heating elements are 120, 230, and 460 volts, with 120 volt the least dangerous. A mold may require high wattage to bring it up to working temperature quickly. A single-phase, 120-volt high-amperage source may not be available. Therefore, a higher voltage may be required. Although any commercial voltage can cause bodily injuries, the lowest voltage offers a better chance of surviving accidental shock. Professionals aware of electrical safety procedures and local codes should design and wire all connections to tooling.

### *Refrigerated Cooling Fluid*

To increase productivity, direct circulation of refrigerated cooling fluid to a mold is sometimes undertaken, but this may cause unforeseen ancillary problems. The chiller unit has a set range from below freezing to approximately 60° F (16° C) with no means of heating the process fluid. If the chiller unit is set with output at a low temperature for a specific machine cyclic speed, any change in pace could upset the mold heat balance. If the thermoforming mold temperature is set too low and removes heat too quickly from a polyolefin plastic (and other resins), formed parts may become warped by the next morning. Forming below optimum temperature can alter wall distribution in a mold cavity and can lock in thermoforming stresses—causing breakage of parts when they are put to use. A chiller should provide low-temperature water to a temperature controller (see Figure 6-9).

### *Water Mineral Deposit Buildup*

Process fluid must be kept clean of impurities and any hard water used should be treated to prevent mineral deposit buildups on mold cooling passages. These passages should be cleaned periodically by circulating acidic compounds to dissolve deposits; even thin coating buildups will decrease cooling efficiency. Severe buildups can reduce the area of fluid passages and restrict flow into cavity locations so that formed parts are not properly cooled.

## Heat Transfer in Metal Molds

Heat transfer through metal mold components follows natural laws. All heat input comes from the plastic web, which transfers heat directly to the cavities after the web is thermoformed. Circulated fluids then extract heat from the mold (see Figure 6-10).

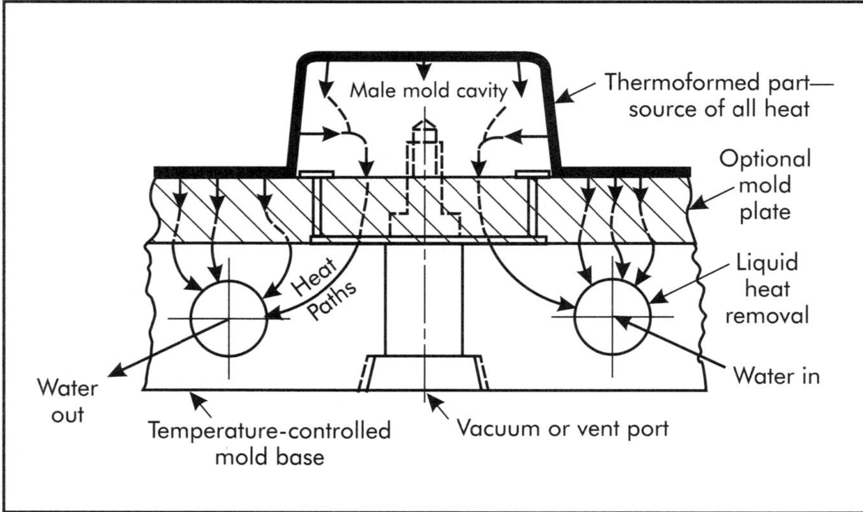


*Figure 6-10. Heat transfer through the mold assembly.*

Heat input must travel the most direct path from cavity to cooling fluid. This heat path is interrupted every time an interface boundary must be bridged, such as at the juncture where two plates fasten together. Heat transfers quite efficiently if two plates being joined are flat, parallel, and fastened properly so that most of their surfaces are in intimate contact. Poor heat transmission takes place when there are air gaps between interfaces, such as when a 0.125 in.-thick (3.18-mm) mold plate is attached to a 2-in. (50.8-mm) thick temperature-controlled plate. Since thinner plates are always slightly warped, they can never be in overall contact with the heavier plate.

## Choice of Mold Metal

Aluminum is the metal of choice for thermoforming mold plates and cavities because it is lightweight, readily available, machines easily, and has high heat conductivity. The coefficients of thermal conductivity for various metals are ranked in Table 6-2, with higher numbers being more efficient.

Aluminum has nearly five times the heat transmission efficiency of steel, but roughly half that of copper. Silver has twice the conductivity of aluminum. Cost and weight considerations eliminate copper, silver, and gold, except for special inserts in cavities. Steel is rust prone, heavy, and a poor heat conductor. Thermoforming does not tend to cause wear to aluminum molds, as is often the case with steel used in injection molds.

Table 6-2. Coefficients of thermal conductivity

| Metal | Heat Conductivity Coefficient ($k$) |
|---|---|
| Steel | 25 |
| Aluminum | 120 |
| Gold | 170 |
| Copper | 220 |
| Silver | 240 |

## High Mold Temperatures

Several plastic resins require the mold to be preheated above the 250° F (121° C) range of water-temperature-control units. These molds can be brought to operating temperature by using electrical heating elements. As soon as thermoforming commences, heat transferred from the hot plastic web often maintains the mold in the optimum temperature range. If the mold temperature falls below the set temperature for any reason, the electrical temperature control instrument automatically switches on the heating elements. When electrically heated molds are used, heat-insulating materials should be mounted to radiating, heat-conducting surfaces not in contact with the plastic web. Thermoforming press platens may require protection from heated molds because hot molds may cause the platens to expand. This expansion can cause jamming of a moving platen. Water cooling of the mold-mounting plate may become necessary.

An estimate of the wattage required to heat the mold cavities and mold base can be calculated by:

$$E = \frac{WS\Delta T}{3.412t} \tag{6-1}$$

where:

$E$ = energy required to heat a mold, watts (J)
$W$ = weight of total mold less the pressure box, lb (kg)
$S$ = specific heat of aluminum = 0.24
$\Delta T$ = temperature change = $T_1$ (maximum mold temperature ° F [° C]) – 68° F (20° C)
$t$ = time in hours required to heat mold from 68° F (20° C) to optimum operating temperature
3.412 = $k$, a constant that converts BTU/hr (heat) into electrical energy in watts

for example:

$W$ = mold weight = 200 lb
$T1$ = maximum mold temperature = 288° F
$\Delta T$ = 288° F – 68° F = 220° F
$t$ = 1 hour

$$E = \frac{WS\Delta T}{3.412t}$$

$$E = \frac{200 \times 0.24 \times 220}{3.412 \times 1} = 3,095 \text{ watts}$$

Since heating is used only for start up and occasionally afterward, a 20% safety margin should suffice.

Safety margin = 20% × 3,095 = 619 watts

Calculating heat loss due to conduction through insulation to the press can be complex; an estimate of 33% heat loss can be used where the mold is properly insulated.

Heat loss = 33% × 3,095 = 1,021 watts

Total energy required = 3,095 + 619 + 1,021 = 4,735 watts

Note: If the heat-up time were reduced to 0.25 hr (15 min), the energy required would be:

$$\frac{3,095}{0.25} = 12,380 \text{ watts}$$

It would be redundant to add a safety margin and a heat loss factor to the 15-min heat-up requirements because more than 2.5 times as much energy has been provided when compared to the 60-min heat-up time. Aluminum melts at 940° F (504° C). It is not unknown for an insulated, electrically heated aluminum mold block to melt if the temperature controller is malfunctioning or incorrectly set. Care should be used during heating.

## Liquid Flow Velocity

Flow velocity through a liquid coolant line is very important because a laminar flow (slow-moving stream) is far less heat-transfer efficient than a turbulent (swift, wild stream) flow (see Figure 6-11).

The flow parameter used to judge whether flow is laminar or turbulent is called the Reynolds number. Some elements that comprise the Reynolds number ($N$) are fluid velocity, density, viscosity, and channel area. The numeric value of this ratio determines

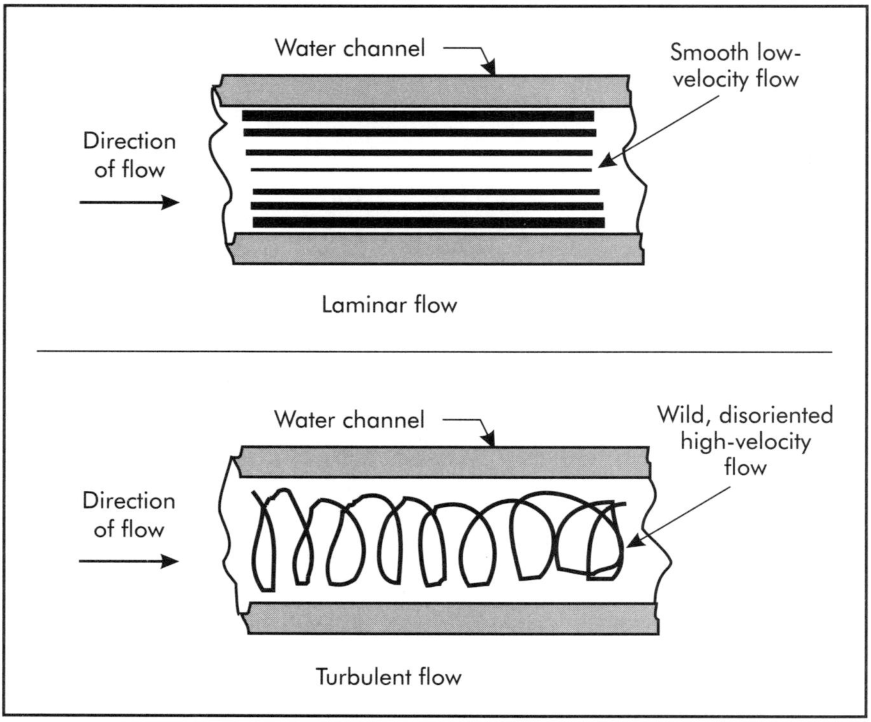


*Figure 6-11. Laminar flow vs. turbulent flow (Bryce 1998).*

the type of flow. When *N* is below 2,000, flow is laminar; between 2,000–4,000, it can be either. Efficient conductivity from mold to fluid requires a turbulent flow with a minimum Reynolds number of 4,000 *N*. A practical flow rate can be obtained by utilizing data from Table 6-3.

Table 6-3. Flow rates (Mitchell 1998)

| Minimum coolant rates for ($N_R$) = 4,000 turbulent flow at 50° F (10° C) | |
|---|---|
| **Coolant Line Inside Diameter, in. (mm)** | **Flow Rate gal/min (L/min)** |
| 0.34 (8.6) | 0.6 (2.3) |
| 0.44 (11.2) | 0.8 (3.0) |
| 0.59 (15.0) | 1.0 (3.8) |

## Distributing Cooling Fluid

Process water on large molds is supplied to a distribution manifold to which supply and return lines are connected (see Figure 6-12). From the manifold, various zones of the mold receive fluid through throttle (variable orifice area) control valves, which can be used to help balance the flow evenly to various mold areas.

The supply inlet and return outlet port-pipe size determine the maximum number of mold service ports that can be accommodated and still maintain turbulent flow, as determined by:

$$P_m = S/M_t \tag{6-2}$$

where:

$P_m$ = maximum number of service ports available for each mold
$S$ = supply inlet pipe inside diameter area, in.$^2$ (mm$^2$)
$M_t$ = total mold service ports × inside diameter area

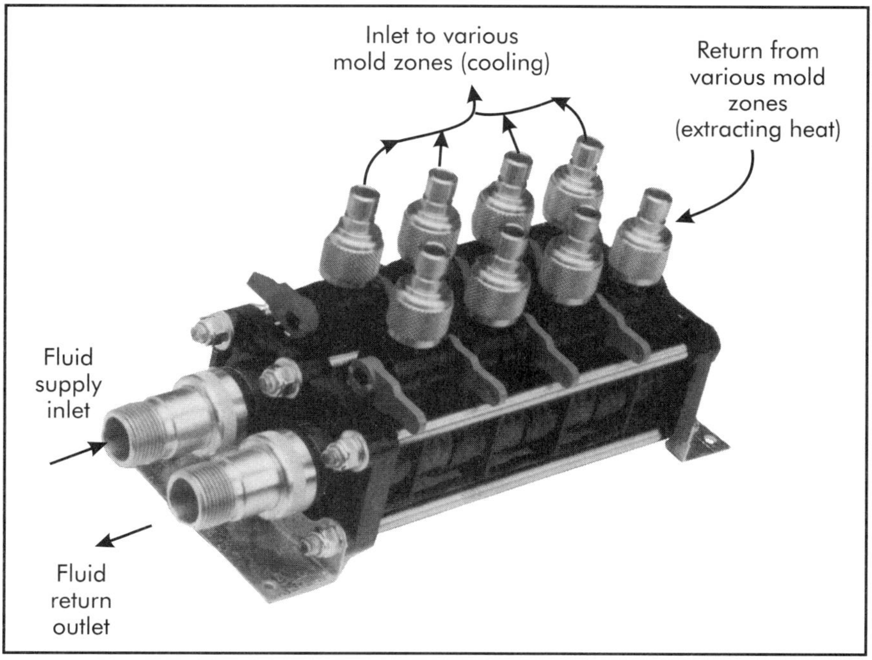


*Figure 6-12. Fluid distribution manifold.*

A fluid path through a mold base can be designed as a series or in parallel and contain a single or multiple loops (see Figure 6-13). For deep molds, flood cooling is used for outside walls of female cavities placed down into a hole. Care should be exercised when flood cooling a mold so that a high Reynolds Number is maintained for maximum heat removal.

A mold base cooled by means of a series path has the fluid path in a continuous loop from the supply inlet port until it leaves through the exit port. Cavities closest to the supply inlet tend to run cooler than other cavities. If the distribution channels and ports are made sufficiently large and the fluid-flow rate is high, the differential in series flow temperature through the mold can provide fairly uniform cooling.

A parallel flow-path configuration has its multiple cooling channels connected internally in the mold base to an inlet and outlet feeder channel. Each cooling channel has a built-in flow restrictor (see Figure 6-13). The restrictor reduces the area in each cooling channel so that the sum of all available open flow areas equals the area of the supply inlet. This device allows fluid flow to be evenly distributed in each channel and the flow remains turbulent. The main difficulty with a parallel path is that the restrictor reduces the flow area radically at its location so that any particles or chemical buildup may change the uniform heat balance at nearby cavity locations and slow down production. The channels can be

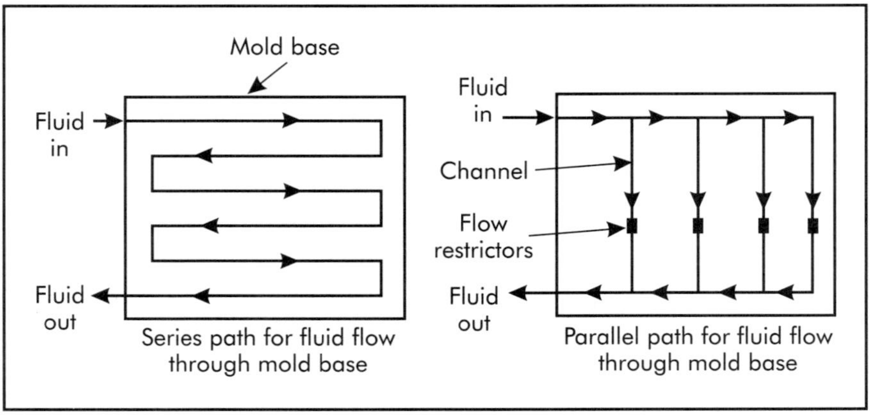


*Figure 6-13. Liquid cooling paths for mold temperature control (Mitchell 1998).*

opened and flushed during maintenance intervals to return heat transfer capability to its original efficiency.

## Clamping of Heated Web

Clamping the heated web to control uniform distribution of part walls over the projected area of the mold is necessary on roll-fed thermoformers. The web, when indexed over the mold, is held and carried by chains along its edges, but is free across its width in the rear and front of the mold. If thermoforming is activated under these conditions, two problems become apparent. As the mold pushes up into the web, rear cavities drag additional hot plastic sheeting from the adjoining oven, causing webs to form on those cavities. The front mold cavities simultaneously pull on previously formed shot, which may be still quite hot, causing the cavities to elongate; that is, round parts can become permanently distorted (elliptical). Front and rear sheet clamps are therefore built into the mold base to prevent these difficulties by isolating the hot sheet's projected mold area (see Figure 6-14). If difficult thermoforming conditions warrant, a four-sided mold clamp may be specified. Such conditions include placing a deep cavity close to the chain edge or processing biaxial-oriented polystyrene (OPS) sheet that can cause the sheet to pull out of the chains due to the plastic's high internal stresses.

Air cylinders that push the clamp forward act as air springs because they are not valve activated, but are supplied continually with regulated air pressure to extend and clamp the plastic sheet as the opposite platen closes on the clamp. Clamps also serve the dual function of assisting ejection of completed shot from the mold when platens are opened after forming.

It is beneficial on some molds to isolate each cavity from the other to prevent one cavity from stealing material from its neighbors—resulting in non-uniform wall thickness from part to part. This method of processing is called *trapped-sheet thermoforming.* This process uses an air-cylinder-loaded plate or an auxiliary press action to clamp the sheet around each cavity before the plugs contact the web. If the cavities have sufficient separation from each other and a four-cavity prototype mold is tested and approved, it

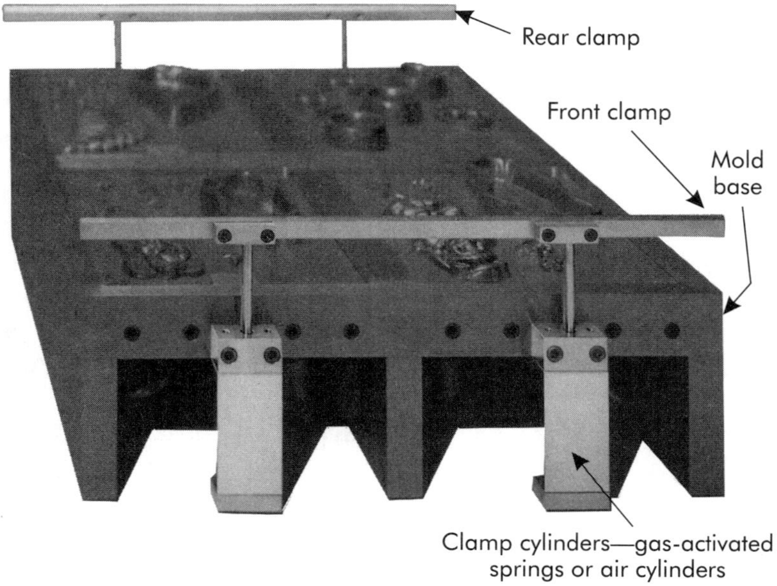


*Figure 6-14. Front and rear clamps. (Courtesy E.D. Segen & Co., LLC)*

may not be necessary to use the trapped-sheet method of forming to prevent sheet stealing.

## Part-ejection Methods

Knockouts and ejection methods are essential for many types of cavities to remove finished thermoformed products. The simplest method used to eject formed parts from a cavity is to release low-pressure air (3–10 psi [21–69 kPa]) back through cavity vent holes or slots while the platens are opening. Front and rear clamps can act to grip the sheet during this process to help lift the shot as the air pressure strips each cavity from the mold. This procedure functions well if there are no severe undercuts and proper draft angles for the wall are provided. Teflon™ coating the mold cavities to reduce friction and adhesion may provide just enough release on difficult cavities to effect part ejection. Teflon provides slippery surfaces, but it also adds an additional barrier to heat transfer and can slow production. Baked-on Teflon coating is difficult to remove from metal if cavity modification later becomes necessary. It is best first to test air ejection on the bare cavity of a prototype mold before Teflon coating a production mold.

## Mechanical Knockouts

Mechanical knockouts are of two types: an ejector plate or knockout pins actuated by a built-in forming press mechanism or driven by air cylinders controlled by an air valve (see Figure 6-15).

An ejector plate is in continual contact with the heated sheet and may require liquid cooling. However, knockout pins are relatively small and are cooled by their contact with the mold. Knockout pins may leave a visible mark on the completed shot, so pin locations should be chosen to blend into the product design. Where mold undercuts are required, the mold may be split at the parting line, which may run through the undercut location to provide assistance in part ejection (see Figure 6-16).

## Thermoforming Foam

Forming plastic foam on sheet-fed thermoforming equipment requires different mold parameters than those for use with solid-plastic-sheet materials. Foam sheet has a very low density when

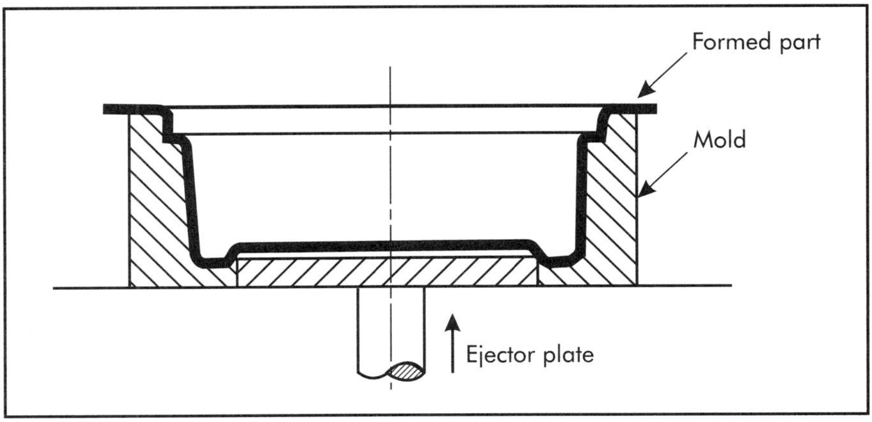


*Figure 6-15. Ejector plate.*

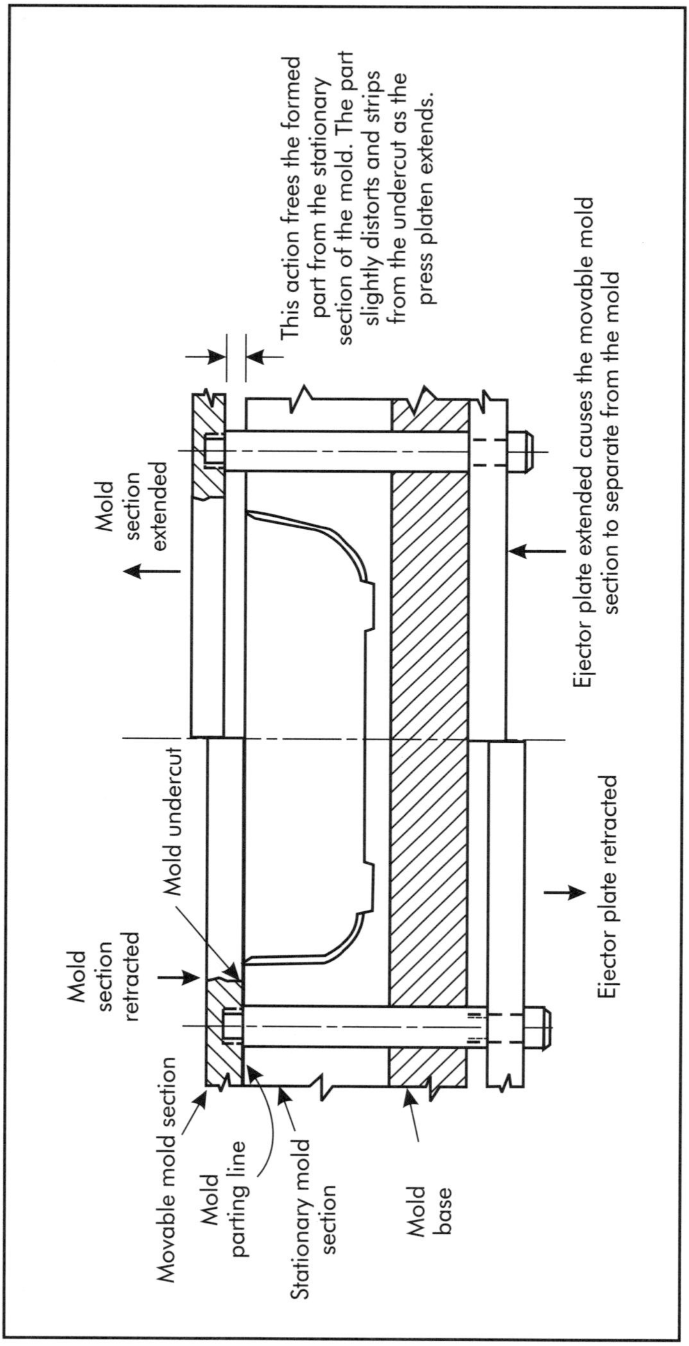


*Figure 6-16. Locating the mold parting line at the base of the undercut helps strip the thermoformed part.*

compared to solid sheet, so its ability to remain heated to proper forming temperature is extremely limited. Processing foam through a thermoforming machine must be very rapid, averaging 20–30 cycles per minute. These materials have limited surface strength. Consequently, they tend to split or break when stressed and forced into deep cavities without generous draft or radii. It is common to specify 5–15° draft with a minimum of 0.50 in. (12.7 mm) radius and to limit the depth of draw ratio (depth of draw/minimum top opening dimension) to less than 0.75.

Foam plastic is formed in molds that have a male half and a conforming female portion with a specified clearance between the closed halves (see Figure 6-17). The clearance need not be uniform throughout the mold since internal entrapped gas blows up the foam plastic to fill the clearance space between the male and female mold segments. The greater the clearance is, the softer the wall (good for cushioning). Reducing wall clearance increases density, thus strengthening the area.

Pressure forming is rarely used with foam, but vacuum forming is sometimes used in very small areas where mold detail (engraving) is required. Forming foam is a mechanical process of the male mold segment pushing heated plastic into the female cavity. The mold should be properly vented to avoid trapping air in cavity pockets, which causes the finished part to be distorted. Not

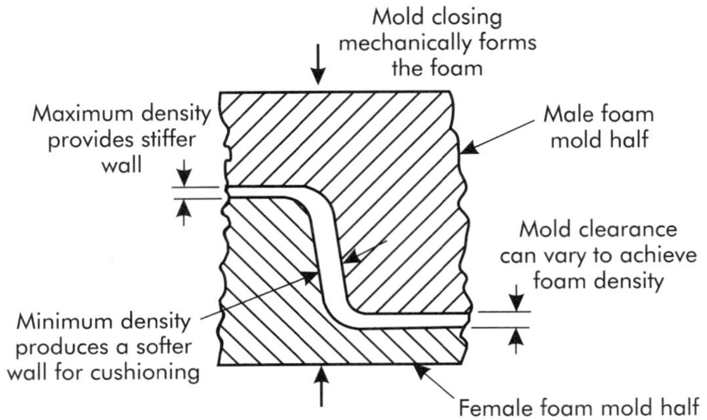


*Figure 6-17. Cross-section of foam mold.*

much can be done to improve the physical strength or appearance of a foam part except to change geometric parameters, which are draft angle, depth, or radii.

Polystyrene foam is specified for its good heat insulating and cushioning qualities and for where it offers a low-weight, less-expensive alternative to solid sheet for throwaway food containers. Polyolefin foams are more expensive, but are sturdier materials that find uses in consumer and industrial products.

## Contact-heat Thermoforming

Contact-heat thermoforming molds are always female in configuration and the process does not allow use of a plug to assist the uniform distribution of wall thickness. This process is favored for use on oriented polystyrene (OPS), but other plastics can be thermoformed with ease. The mold is mounted to the top moving press platen and located inside and below the edge of a knife-like die (see Figure 6-18). The die can be the steel-rule type, but is

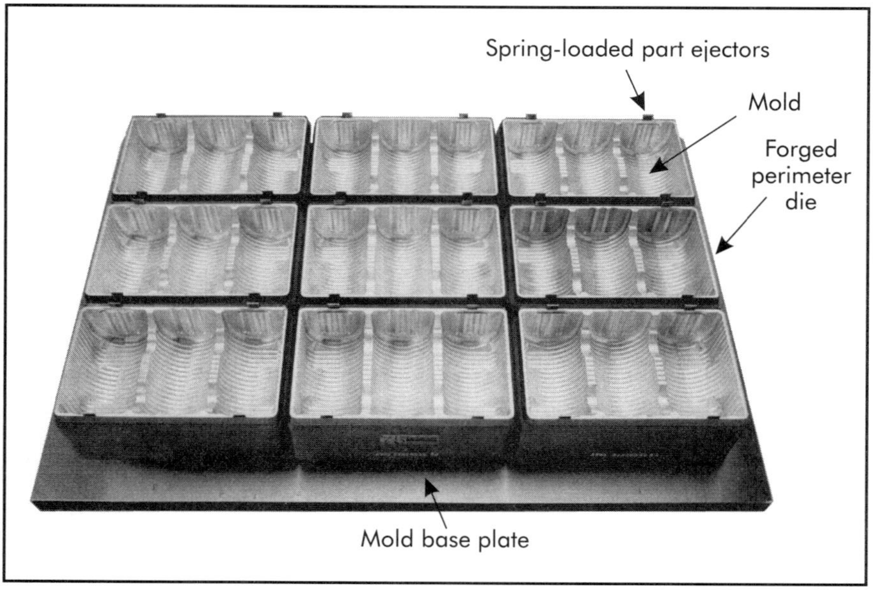


*Figure 6-18. A mold mounted to a top moving press platen.*

preferably a forged or machined tool-steel die, which provides increased knife life. When the formed plastic component calls for a minimum flange, the wall of the die near its knife-edge can serve as part of the mold sidewall. Since this forming process requires that the finished plastic part be cut in place and held to the web by weak, partially cut bridging nicks, mold undercuts may cause these tabs to break during ejection. The formed part can be cleanly stripped from the cavities as the mold is opened if the finished plastic components have a sufficiently wide flange to place a series of spring-loaded devices near the undercut.

## Mold Temperature Control

Temperature control of contact-heat female molds is most often provided by directly fastening them to an aluminum base plate that contains circulating fluid channels. When the mold is deeper than 2 in. (50.8 mm), heat transfer from the mouth of the mold to its base may become inefficient and the process can slow and become uneconomical. Liquid flow circulating up from the mold base through channels in the mold cavity may be necessary to provide proper mold temperature control of deep molds.

## Accuracy of Mold Buildup

It is critical to maintain proper thickness and flatness control of all dies and plates used to fabricate a contact-heat mold. Dies are designed to cut part way through the sheet thickness when they are closed on the plastic to initiate the process. This initial contact with the sheet allows the die to seal off the mold, thereby heating entrapped plastic and building pressure in the female mold cavity. Final trimming takes place when the secondary stroke of the press extends further to cut through the remaining sheet thickness. If a multi-cavity mold-and-die assembly is not flat and parallel, it may cut completely through the sheet at various points rather than part way. Cutting completely through the sheet during heating or forming results in material trapped inside the mold being improperly heated and not formed. The cutting stroke of the press acting on an inaccurate die may result in incomplete trimming of formed parts, which may not allow clean stripping from the web.

Setup personnel may adjust a contact-heat tool for small accuracy errors when it is being fitted into its press. Even small accuracy discrepancies can cause laborious stoning or shimming, which is time consuming and costly. Careful inspection before storage of the tool may avoid delay when the mold is next installed in the machine.

## Conclusion

Before proceeding with design of the mold, it is best to choose among competing options and form a priority list of factors that shape mold design. Almost all these factors should be weighed against one another. Usually, the available time for mold completion is preset and is always too short. Tool price requires constant vigilance to choose the most effective design and fabrication technique. Long-term delivery of purchased materials and mold components can be arranged as soon as the need becomes apparent. Prototype sample parts of the highest quality are essential to proceed with confidence in design of a production mold. The designer should not take such ownership in his creation that he closes his ears and mind to useful suggestions or modifications. A design is only an idea presented on paper. An eraser or computer mouse can update the design to avoid expensive rework in the mold shop. On the other hand, the design should not be judged or changed by every breeze blowing through the office.

## References

Bryce, Douglas. 1998. *Plastic Injection Molding...Mold Design and Construction Fundamentals*. Volume III: Fundamentals of Injection Molding series. Dearborn, MI: Society of Manufacturing Engineers.

Mitchell, Philip, ed. 1998. *Tool and Manufacturing Engineers Handbook*, Fourth Edition, *Volume 8: Plastic Part Manufacturing*. Dearborn, MI: Society of Manufacturing Engineers.

# 7

# Mold System Components

The mold-base assembly is essentially the same whether operating with pressure or vacuum forming, or using male or female cavities. A pressure box is a required component in pressure forming. When vacuum forming, a vacuum seal-off box is used to seal the mold. Plugs, assists, and other auxiliary techniques distribute the wall thickness more uniformly over the body of the thermoformed parts using either process. Family-style tooling, which allows a variety of cavities to be run on a single mold assembly, and other options, may reduce tooling costs.

## Mold Base Assembly Components

The parts for a mold base assembly are the base, cavities, sheet clamps, and pressure box components.

### Mold Bases

Mold bases are temperature-controlled plates to which the cavities transmit heat they receive from the hot plastic web. The base contains a port to vent the mold cavities. The port also can introduce a vacuum during the forming process (see Figure 7-1).

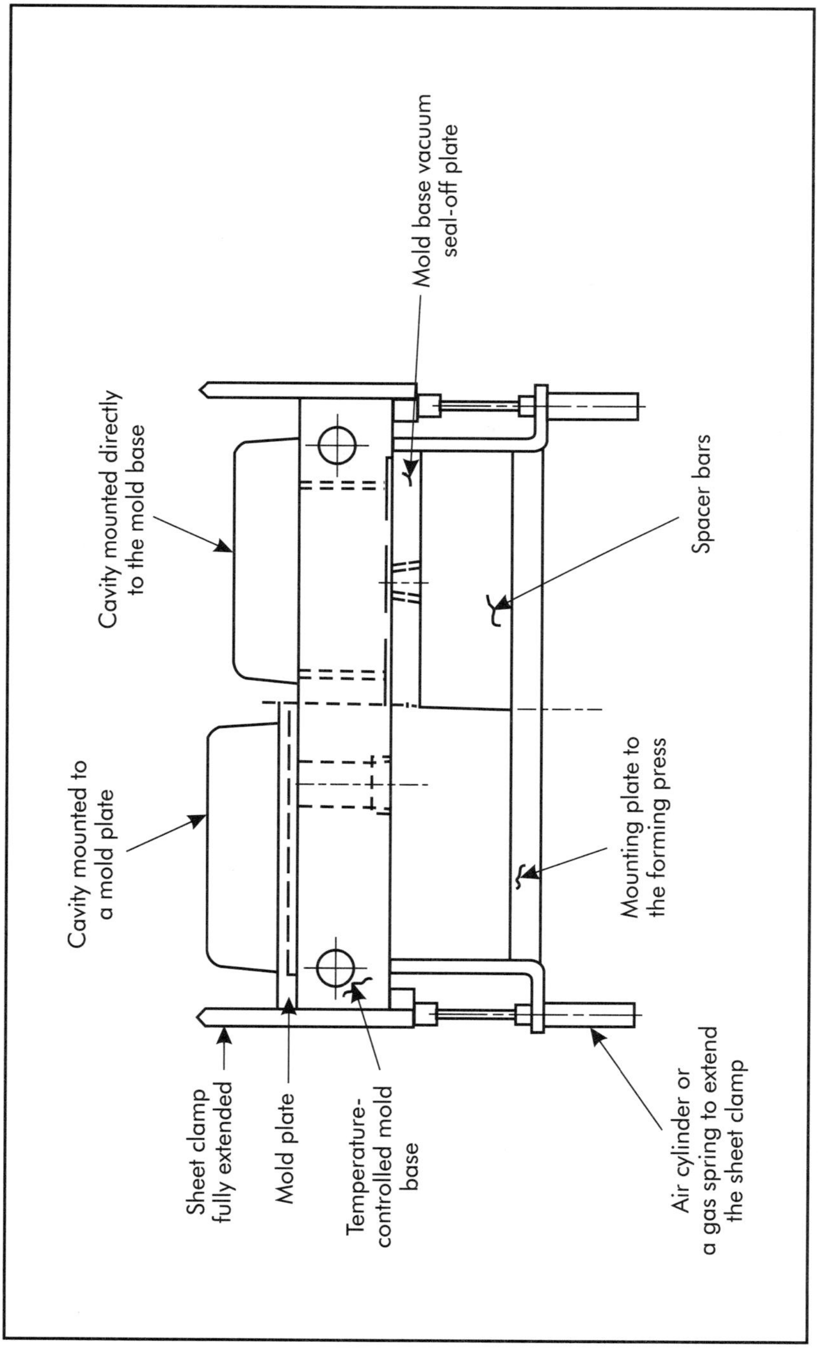


*Figure 7-1. Mold base assembly.*

## *Cavities*

Cavities can be fastened to a mold plate and this completed assembly can be mounted as a single unit to the mold base. Different cavity assemblies using the same size mold plate can be installed on a mold base of equal dimensions. Alternatively, cavities can be fastened directly to the temperature-controlled mold base for more efficient heat transfer. This style of mold base can only serve other cavities that have a similar footprint. When pressure forming, the mold is enclosed by a pressure box, which is mounted on the opposite moving platen and is later filled with compressed air after the tool halves are closed. When vacuum forming, the earth's atmospheric pressure acts on the heated sheet as a vacuum is drawn on the opposite side of the web. A seal-off box prevents any vacuum leakage. The spacers and mold mounting plate serve the same function, locating tooling halves at the sheet line for vacuum forming as well as pressure forming (see Figure 7-1).

Air cylinders or gas springs power sheet clamps (front and rear) and act with nearly uniform force during thermoforming. The clamps are fully extended in their normal operating position. When the platens close, the clamps retract away from the platens reacting against them. The clamps extend when the platens open, helping eject the finished part.

## *Sheet Clamps*

Sheet clamps are attached to the mold base when male cavities are being thermoformed. During the thermoforming cycle, the pressure- or vacuum-seal-off box extends first to the sheet line. As the mold base extends toward the box, the clamps entrap the sheet to ensure consistent forming conditions.

Female cavity thermoforming calls for the female mold assembly to extend first to the sheet line. Then the pressure- or vacuum-seal-off box, with its plugs, follows to seal the web. Sheet clamps function best for female cavities when attached to the pressure box. They securely grip the sheet before the plugs start to enter the cavities, prevent distorting any completed, partially heated parts, and avoid drawing surplus sheet from the oven.

## Pressure Boxes

Pressure boxes are designed to contain internal air pressures of at least two times the maximum plant airline pressure. The box should include a safety valve set for approximately 50 psi (345 kPa) and a conforming groove to coincide with front and rear mold-base sheet clamps. A silicone rubber-sealing gasket runs completely around the perimeter of the box opening to seal the pressure-forming air within the box (see Figure 7-2).

The silicone rubber seal is contained by friction in a snugly fitted groove. The seal is occasionally held in place with silicone rubber cement, but this type of adhesive requires several hours to cure and can cause excessive production downtime if it needs replacement during a run. The cross-section of the seal can be round (as an O-ring), rectangular with relief grooves in each rectangular face, or an "A" cord (see Figure 7-3). The cord ends are skived to prevent air leakage at the end joints (see Figure 7-4).

The height of pressure-box walls varies depending on whether the matching cavity is male or female (see Figures 7-5 and 7-6). The pressure-box depth should be equivalent to the male cavity height plus a 0.25 in. (6.4 mm) minimum clearance under the baffle plate. The female mold needs only a minimal pressure box clearance of 0.50 in. (12.7 mm) above the sheet to be formed, plus space

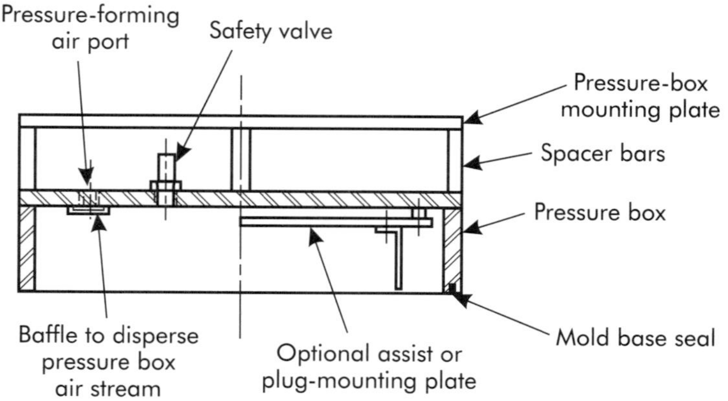


*Figure 7-2. Pressure box assembly.*

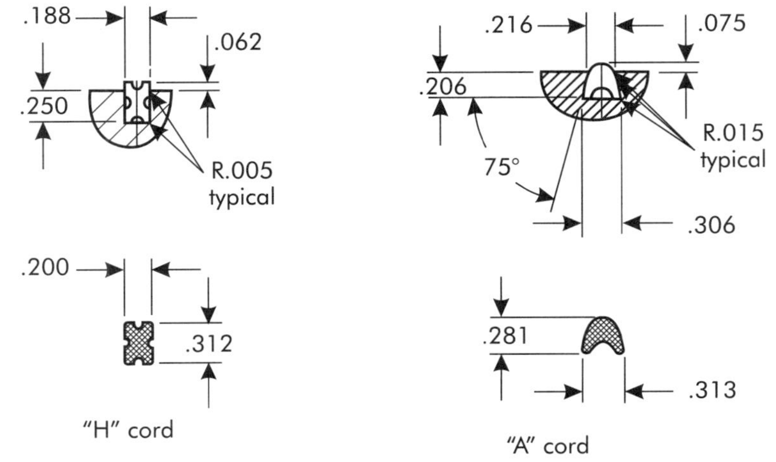


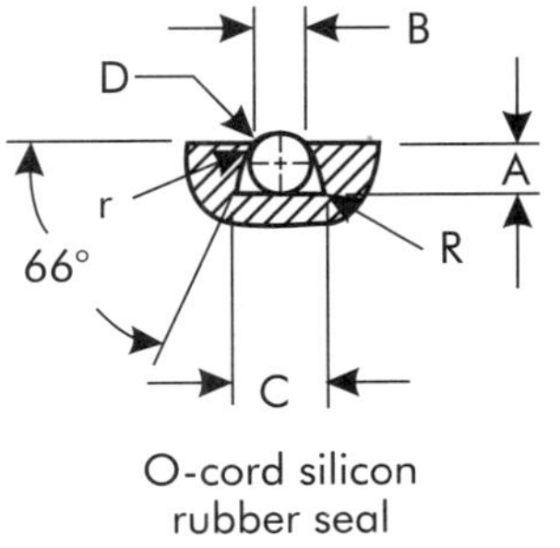


| D | A ±.001 | B ±.002 | C | R | r |
|---|---|---|---|---|---|
| 1/8 | .101 | .103 | .193 | .032 | .010 |
| 3/16 | .151 | .155 | .290 | .032 | .010 |
| 1/4 | .201 | .206 | .385 | .063 | .015 |
| 5/16 | .260 | .250 | .481 | .094 | .020 |

* Measurements in U.S. customary units

*Figure 7-3. Typical pressure-box seals. (Courtesy E.D. Segen & Co., LLC)*

for any male projections that are above the female cavity. The pressure-box internal air volume often can be greater than the combined volume of all the air cylinders used to run a thermoforming machine. The depth of the walls should be kept to a minimum. Reducing the volume of an existing pressure box through use of non-flaking material to fill unused space can reduce the pressurized air required for each cycle.

The female-mold pressure box consists of a thick plate with a machined pocket conforming to the mold area and deep enough to clear mold projections. Male-mold pressure-box walls are made

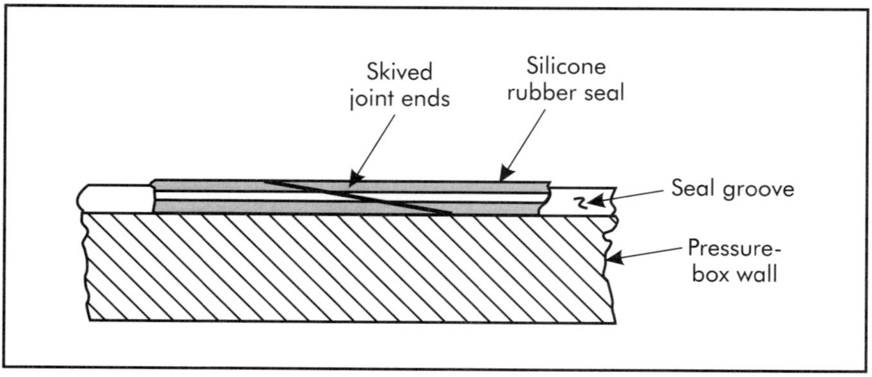


*Figure 7-4. Seal-joint ends are skived to prevent pressure-box leaks.*

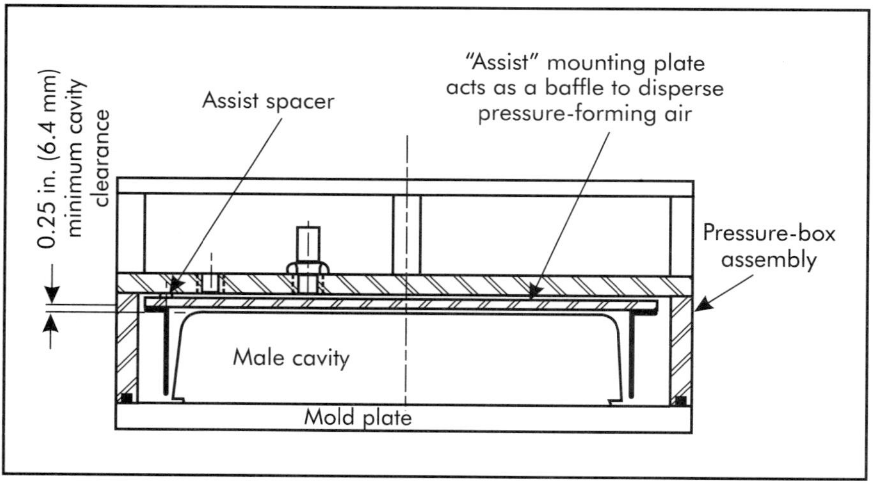


*Figure 7-5. Clearance from the top of the male cavity to the assist plate.*

of metal bars and securely fastened to a metal base plate. The walls are further secured by dowel or roll pins. Welding is not recommended because of the danger of incomplete fusion at critical junctions. All bolted metal joints can be sealed with silicone rubber cement to prevent air leakage.

A baffle should be provided at the pressure-forming entrance port to prevent the jet of air from impinging directly on the hot sheet, which can freeze and mark the plastic in that area. When the air jet is directed at the baffle, the air stream is split and moves

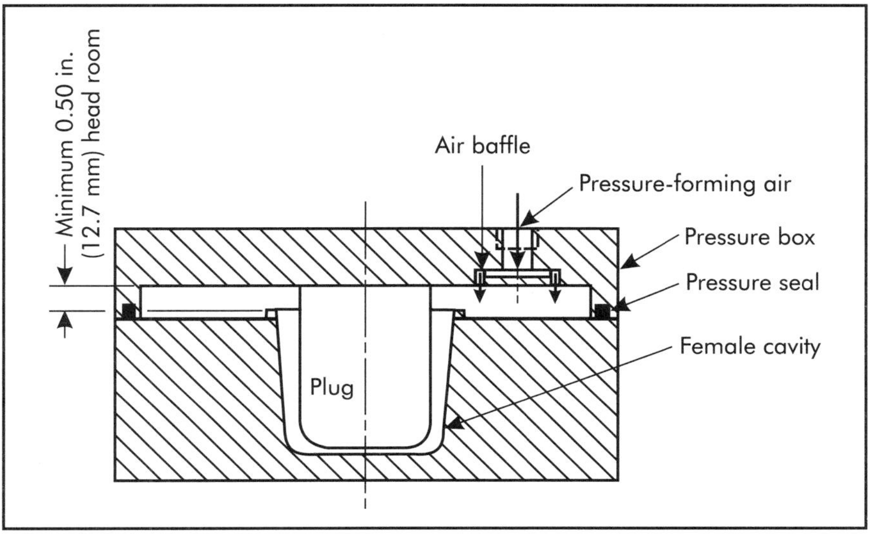


*Figure 7-6. Female cavity plug and the pressure box.*

harmlessly at right angles from the inlet port. A large baffle plate can also serve as a platform to mount either assists or plugs aligned to the cavity centerlines.

## Spacers

Spacers are attached to the back of the pressure box to adjust the varying depth of the pressure box to the relatively inflexible shut-height of the forming press. The spacers are constructed of rectangular-shaped bars approximately 1 in. (25.4 mm) thick and are set 8–12 in. (203.2–304.8 mm) apart for a uniform distribution of press locking pressure on the pressure box. This construction allows an open area for air and electrical lines to be installed to the pressure box.

## Mounting Platen

The pressure-box mounting plate is bolted to the top face of the spacers and contains mounting holes, which attach the whole assembly to the forming-press platens. Many press platens contain keyways located on their *X* and *Y* centerlines to line up the pressure-box assembly to the corresponding mold assembly mounted on the opposing platen. The pressure-box mounting plate can be designed with similar keyways to which keys can be permanently fastened for rapid and accurate tooling setup.

### *Clamp Bars*

Platen clamp bars are necessary to secure the complete assembly to the press platen on machines with open-grillwork platens, which are designed to reduce platen weight yet retain structural stiffness. Other types of press platens are fabricated from solid plates that have bolt-mounting holes and do not require clamp bars.

### Adjustable-length Mold Bases and Pressure Boxes

For a given web width, adjustable-length mold bases and adjustable-length pressure boxes can be specified and reused with different length mold assemblies. Adjustable-length tooling cost can be distributed among many individual projects.

Adjustable mold bases have a main section of fixed length with loose segments of 1 in. (25.4 mm) increments, up to a collective additional buildup length of 7–8 in. (177.8–203.2 mm) (see Figure 7-7). All loose segments can be bolted together to the main section and are provided with liquid-temperature-control passages. The adjustable mold base requires cavities to be mounted on a mold plate; this assembly is then fastened to the adjustable mold base. The back sheet clamp is permanently mounted to the mold base and the front clamp assembly is attached flush to the front end of the last base segment. Theoretically, any number of loose segments can be added to an adjustable mold base, but a large number of segments become impractical to manage in the real world of production.

If the customer can accept 15–25% reduced thermal efficiency, the width of the mold plate can be extended to overhang up to 1 in. (25.4 mm) on each adjustable mold base to provide additional

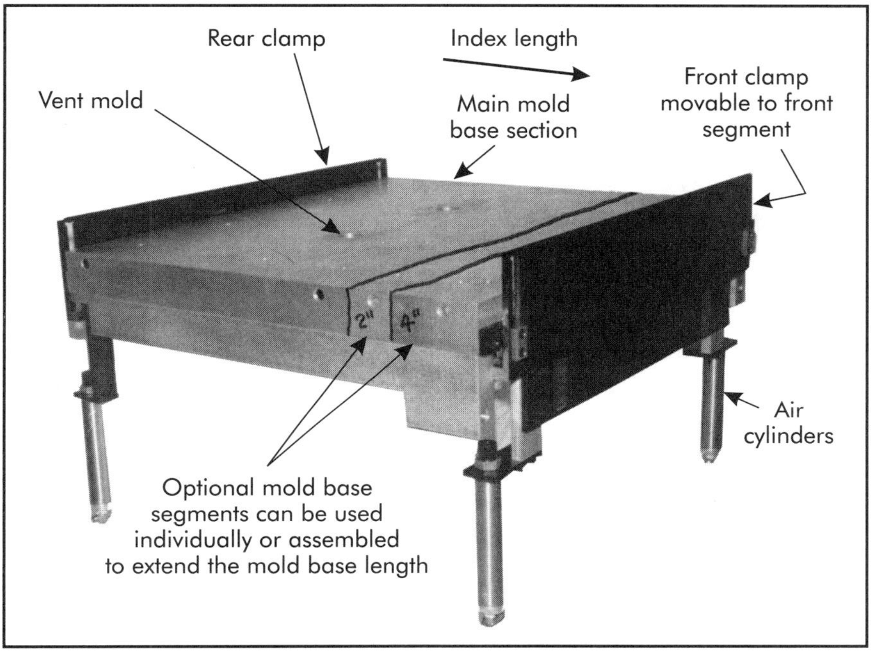


*Figure 7-7. Adjustable-length mold base.*

versatility for this component. Adjustable, multi-segment mold bases are never as thermally efficient as a single, monolithic base and require additional time to set up. The loose segments of machined aluminum tend to become nicked and damaged during handling. In addition, the correct length fasteners are often mislaid and may delay machine setup.

Adjustable-length pressure boxes are in common use at custom thermoforming shops that handle a large variety of small-volume jobs (see Figure 7-8). The loose front side panel of the adjustable-length box can be positioned in a machined slot from the longest to the shortest adjustable design length. The three fixed sides are permanently attached to the top pressure-box plate and the front panel is bolted into the desired position to complete the box. Care must be exercised in designing adjustable-length pressure boxes to provide maximum safety to operators. On the shop floor, tool assembly safety becomes the responsibility of often-untrained installers who may not fully understand proper assembly details.

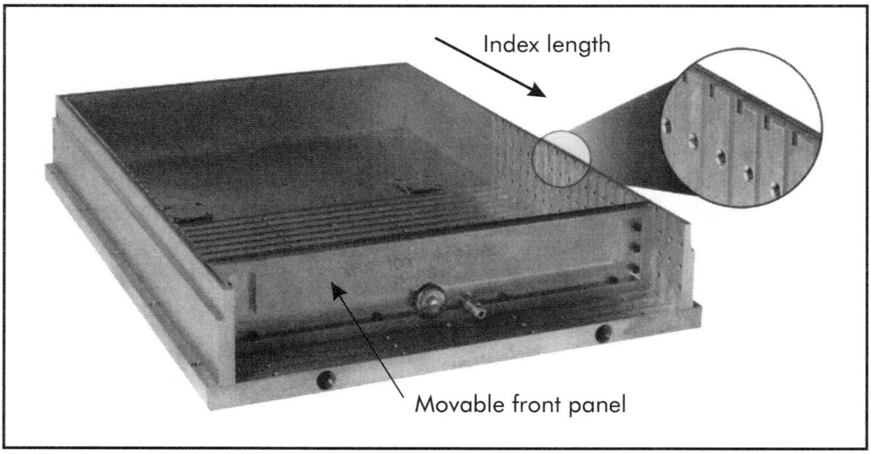


*Figure 7-8. Adjustable-length pressure box. (Courtesy E.D. Segen & Co., LLC)*

## Vacuum-forming Seal-off Methods

Pressure boxes are built to withstand high internal pressures and often are too costly a production tool for low-volume, uncomplicated thermoformed parts. In that case, vacuum forming is the natural alternative, using a simple vacuum seal-off box. Any mold assembly designed for pressure forming can be used for vacuum forming. Vacuum forming requires only a tight vacuum seal on each of the four edges of the mold plate and a means of mounting thermoforming plugs and assists.

An adjustable vacuum-seal-off box vented to the atmosphere and built of light aluminum sidewalls is practical for various length molds (see Figure 7-9). Seal-off boxes are suitable for roll-fed thermoforming machines where the sheet is held in the front and back by air clamps. The seal-off box is mounted to the platen opposite the mold, which provides a complete perimeter seal to the mold plate.

Vacuum forming is not well suited to cavities requiring sharp detail or uniform undercuts. It tends to generate more imperfectly formed shots due to its low inherent forming pressure and can be a slower process than pressure forming because part cooling is less efficient. These deficiencies must be measured against tooling cost savings when thermoforming small-volume production jobs.

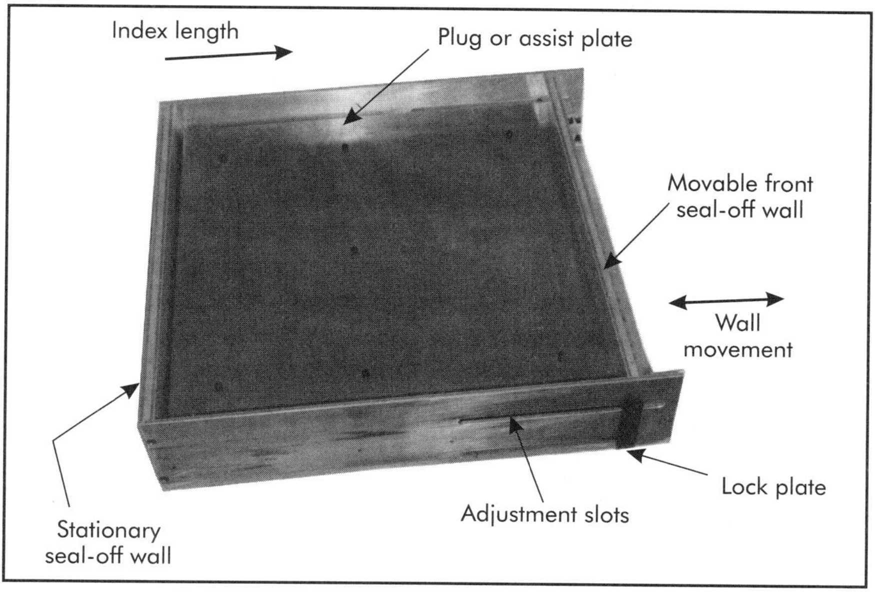


*Figure 7-9. Adjustable-length vacuum-forming seal-off box.*

## Family-style Tooling

A group of thermoformed products having the same exterior dimensions allows interchangeable individual cavities to share a common master mold, pressure box, and trim dies for products with different depths or styling. Cavities having the same outer configuration may be designed into a family tool protocol, even though cavity interior layouts may vary from one to another. Varying food trays are made with the same tooling systems, using interchangeable cavities with a variety of separate compartments.

A family of female cavities may require completely different plug geometries to achieve a uniform wall thickness for each varying product in the tool family. Male family cavities may need separate assists for each individual product. This versatility of mix-and-match products within one mold requires creative ingenuity to design plugs and assists that are simple to install on the production floor. The family's trim tooling can use the same perimeter dies, but may require individual locating pilots for each of the products.

Other types of family tools may be specified when the customer is willing to accept a less-efficient cavity layout for the ability to incorporate a group of associated parts of different shapes that may be required to fill an order. For example, a master mold might include a set of cavities for producing drinking cups varying in diameter from 2.25–3.50 in. (57.2–88.9 mm). One master mold base and die could be used to produce cups with six different diameters. Knowledge of potential quantities of different cup sizes to be produced is the basis for grouping related cavities to make the best use of the available mold area. If 60% of the product is in the 2.25–2.63 in. (57.2–66.8 mm) range, then the mold base area is designed to interchange cavities within this group. The other odd-sized cavities share the space remaining. The trim dies required for this family mold mandate that each mold cavity lane, front to back in the direction of index, be of the same diameter and style to avoid mixing product types. Economical use of the sheet area precludes too wide a variation in die diameter of interchangeable inserts. A difference in area of 2.25–2.63 in. diameter (57.2–66.8 mm) for cups produces 52% additional waste when compared to building a separate mold for each diameter of cup.

Family-style tooling has other trade-offs. It reduces initial investment cost; however, it raises the piece price for resultant thermoformed parts, such as cups, in several ways, including:

- Sheet area is wasted when the smallest cups are run in an area designed for larger cups.
- The sheet thickness specified to form a family of drinking cups requires that the largest and smallest cups be thermoformed from the same web thickness. The resultant smaller-diameter cups are overweight (more expensive) to accommodate the largest diameter cups in the same web.
- The ratio of sizes required for any given production order might not be in harmony with the ratio designed into the tooling. This situation could cause an overrun of one or more cup sizes to produce the required quantity for one size on order. Also, spoilage of some thermoformed parts in one lane of cavities can cause an imbalance in production quantities, which will cause a further excess of unwanted products. It is highly unlikely that an odd batch of cups could be sold. Therefore, they would be reground into recyclable scrap.

- Handling and collating variously sized parts trimmed from a web can increase labor costs and other expenses at the machine take-off point. If any process spoilage has occurred, completed parts may require hand sorting to balance out quantities for delivery.

## Design of Plugs and Assists

A *plug* or an *assist* is a mechanical aid for distributing hot plastic before the vacuum- or pressure-forming valves have been activated. Mechanical forces that stretch the plastic web enhance or inhibit slippage of plastic as it is displaced. Prominent factors include the geometry of a mold plug or mold assist, and heat loss from sheet to plug or assist. Secondary factors may include the coefficient of friction or the size of the radius in contact areas between a moving plug or assist and the plastic web. Some formed parts have both male and female segments, so a combination of a plug and an assist may be required.

### Plug Geometry

The geometry of plugs and assists can be developed using computer software available from a number of sources. Alternatively, these items can be designed by rule-of-thumb criteria and refined in prototypes. It is imperative that the plug modification be done during the sampling phase since a multi-cavity production mold could require remachining many plugs and/or assists, which could tie up a production thermoformer indefinitely (see Figures 7-10, 7-11, and 7-12).

### Plug Design

When designing a plug, the initial sizing should contain sufficient extra material to allow slowly machining away the shape as dictated by thermoforming sampling. Preliminary plug length should extend to within 0.12–0.19 in. (3.0–4.8 mm) clearance of

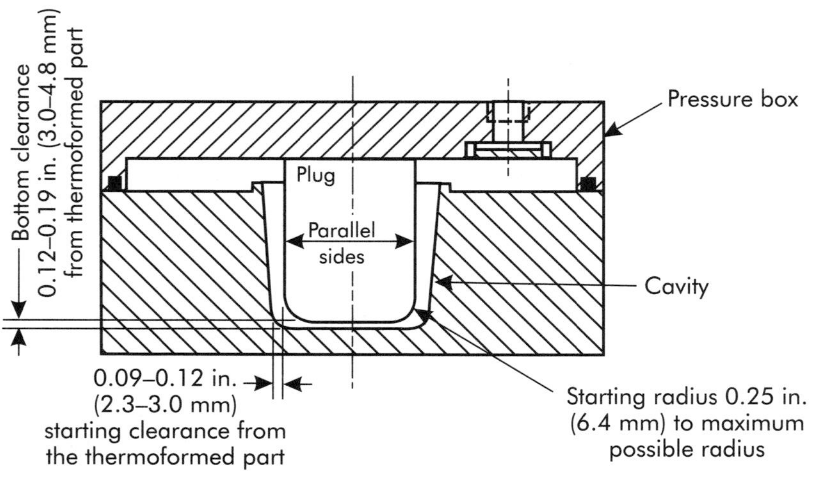


*Figure 7-10. Initial plug design parameters.*

the bottom of the plastic in the mold. The sides of a plug need not follow the draft angle of the cavity but can be straight and parallel to each other. A clearance of 0.09–0.12 in. (2.3–3.0 mm) per side of the formed part at the minimum mold dimension should provide a good starting point for a plug cross-section.

Determining the plug-tip radius size requires some trial and error and can be best developed while sampling prototype parts. If the radius is too small, hot plastic material will not roll easily off the tip to help build up component sidewalls. Specifying too large a tip radius can prevent sufficient plastic from flowing to the area where the cavity base joins the sidewalls. A starting tip radius of 0.25 in. (6.4 mm), depending on the cavity area, depth of draw, and geometric shape of the finished part, may be specified, but a larger radius may become necessary.

## Assists

Assists are devices used to uniformly distribute hot sheet along the sidewalls and flanges of male molds (see Figure 7-11). They are mounted on the inside of pressure boxes, allowing for 0.06 in.

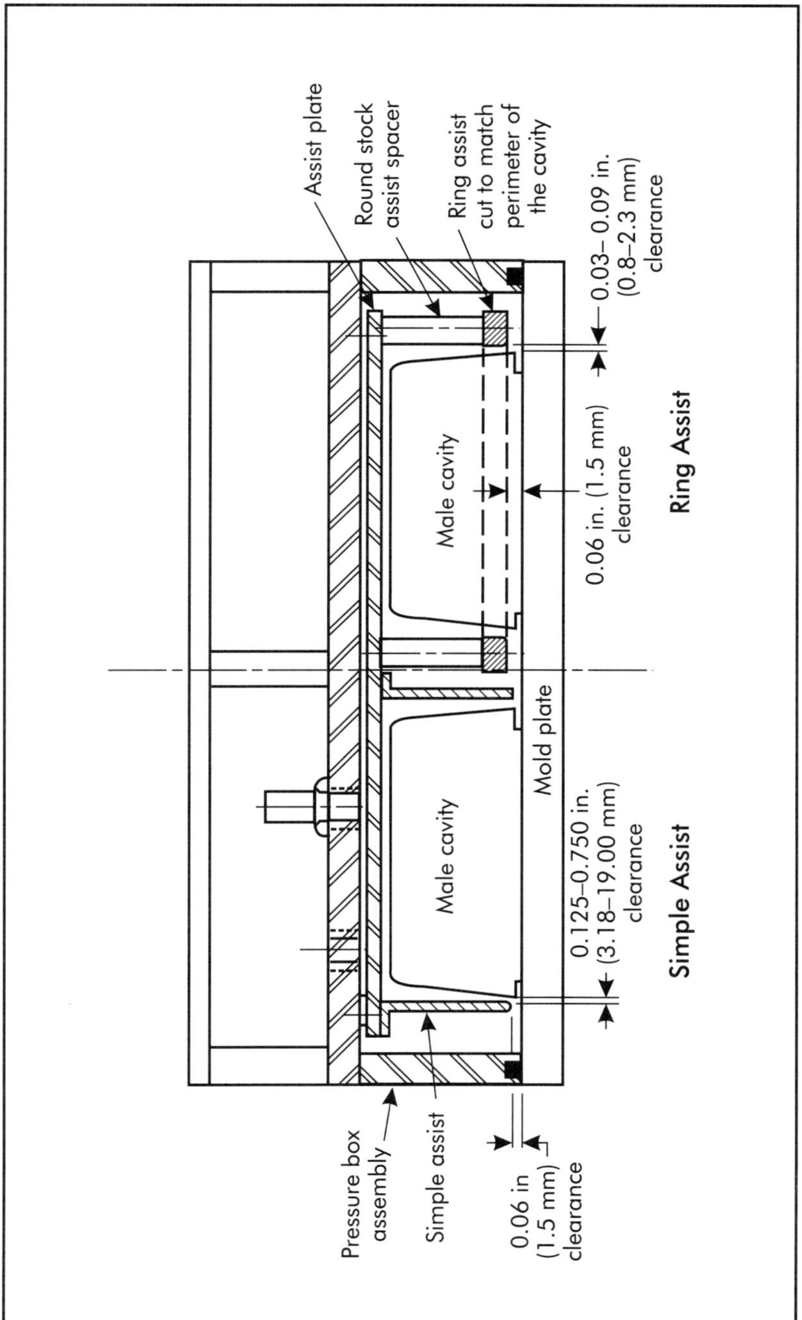


*Figure 7-11. Types of male mold assists.*

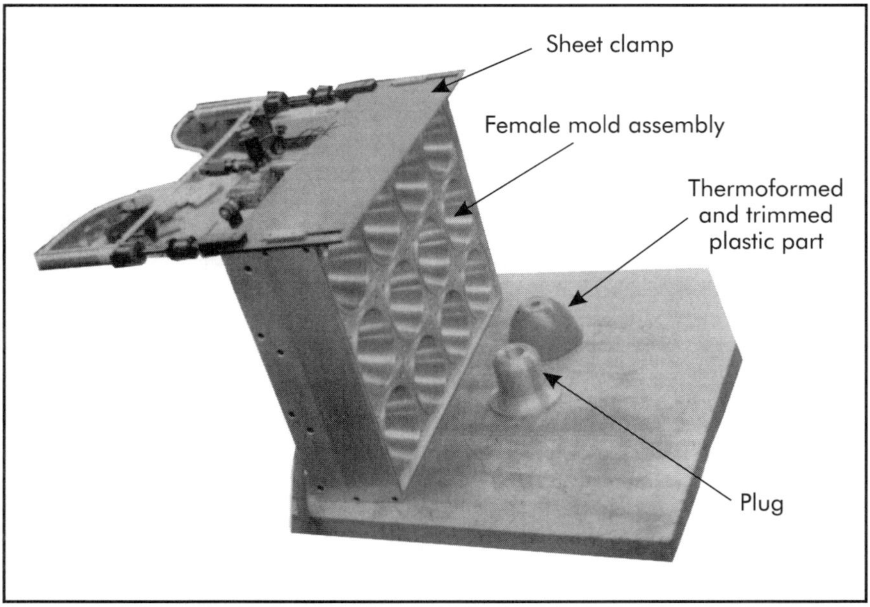


*Figure 7-12. Multiple-cavity female mold plug and finished plastic part.*

(1.5 mm) clearance above the flange sheet thickness after forming. Sidewall clearance can be 0.03–0.09 in. (0.8–2.3 mm) per side for a ring assist after allowing for sheet thickness, and 0.125–0.750 in. (3.18–19.05 mm) clearance for a simple assist.

## Heat Loss to Plug

Heat loss from sheet to plug or assist causes plastic to cool at all points in contact with the moving plug or assist. This cooling often prevents the plastic flow from producing a uniform wall thickness. Heat transfer during this process can be minimized either by raising the temperature of the plug or assist or by specifying an insulated plug or assist material. The surface temperature of the plug or assist should be brought close to that of the heated web so that very little heat loss takes place.

Raising the temperature of a metal plug by means of an electrical heating element requires close monitoring by a sensing thermocouple, located close to the tip and controlled by a temperature-regulating instrument. If the plug temperature rises above that of the sheet, it will burn through. If it is too cold, the plastic will chill. Certain types of plastics present a very narrow heat-window range in which the plug will operate; perhaps a maximum of 7–10° F (4–6° C) for polypropylene, with a much broader range for polystyrene, high-density polyethylene, and polyvinyl chloride (PVC). If a plug is overheated, it may become fouled with molten plastic. Teflon™ coating can reduce cleanup time. Teflon may also lower the coefficient of friction, allowing the sheet to slide more easily over a plug.

An electrically heated plug is essential for prototype thermoforming because the process is so inherently slow that even a plug fabricated of insulating material will not retain sufficient heat to maintain a stable forming temperature. High-speed thermoforming cycling allows acceptable production to proceed after 10 or so scrap shots build up sufficient surface heat on an insulated plug.

## *Insulated Plugs*

Insulated plugs have been fabricated of epoxy, wood covered with felt, or syntactic composite foam, which contains mini-glass-balloon air cells in a low-conductivity matrix. Syntactic foam is a fairly expensive raw material, but its use tends to avoid pitfalls found with other insulating materials. Wood may dry out and split from the heat; felt can leave residual fibers in food or medical packaging; and epoxy takes a long time to reach proper operating temperature. Insulated plugs work well but must be reheated by running scrap shots every time a production machine is shut down.

Electrically heated metal plugs for use on multi-cavity molds are very efficient, but add to tooling costs and complications that must be balanced against the plastic waste incurred when using insulated plugs.

## *Assist Materials*

Although assists function more efficiently if made of insulated material, aluminum is generally specified because of its greater strength and durability. Occasionally, an operator closes the platens and tooling on a cold plastic sheet. This can damage the light construction of the assist if it is not fabricated of metal. Normally, the assist is brought up to operating temperature by its contact with the hot web and the first few shots are scrapped until the proper assist temperature is maintained. All assist contact edges should be radiused and polished to allow plastic to flow freely along the cavity sides.

## Other Thermoforming Systems

There are other processes used to thermoform some products such as contact heat and foam thermoforming, trapped sheet forming, coining, and creasing. Specialized methods are used to produce vents in hot food containers, drinking straw openings, and screw-type lids.

### Contact Heat and Foam Thermoforming

Plugs and assists cannot be used in some circumstances. For instance, foam molds or contact-heat pressure-forming tooling do not offer any means of utilizing plugs or assists. Generous radii, sidewall draft, and heavier sheet thickness can control wall thickness in these types of molds.

### Trapped Sheet Forming

Trapped-sheet forming of female molds offers the feature of isolating each cavity by clamping the outer perimeter before the plug is activated. It is claimed that, when using this technique, material is not stolen by nearby plugs as they enter the cavities. European-style, trim-in-place thermoform tooling incorporates the trapped-sheet technique as part of the process. Many U.S.-built thermoforming machines may require the trapped-sheet feature to be built into each mold using auxiliary air cylinders to activate the trapped-sheet plate. The merit of using this process is based on the quantity of parts to be produced to amortize the added costs compared to additional sheet weight needed when cavities require more separation or when sheet thickness is increased to produce an equivalent product.

## Coining

*Coining* is the mechanical compression of plastic that causes it to flow under pressure between the mold and a heavy, spring-loaded coining tool mounted in the pressure box. This process is used for critical thickness areas, such as flange portions of containers to be sealed with metal lids, or to emboss decorative detail into portions of a part. Displacing plastic when the sheet is still hot requires far less pressure than when it is at room temperature. Occasionally, coining is used to maintain a uniform flange required for radio-frequency heat sealing. The sheet also can be coined by means of auxiliary cylinders transmitting pressure through a plate to flange-sizing inserts before plugs touch the sheet. Coining may require calculating press lockup force to prevent the press from opening and venting its air pressure. This calculation involves establishing lockup force minus both the opening force of pressure forming and the force of coining springs or air cylinders.

The maximum allowable coining force during thermoforming is found by:

$$F_C = (F_L - F_O)\,E \tag{7-1}$$

where:

$F_C$ = maximum allowable coining force, lbf (kN)
$F_L$ = machine manufacturer's thermoforming press lockup force specification, lbf (kN)
$F_O$ = area of pressure box, in.$^2$ (m$^2$) × forming air pressure, psi (kPa), = opening force, lbf (kN)
$E$ = estimated percentage available for coining due to friction and other losses = 80%

Therefore:

$$F_C = (F_L - F_O)\,80\%$$

For example:

$F_L$ = 40,000 lbf (178 kN) lockup force
$F_O$ = 25 × 20 = 500 in.$^2$ (0.3 m$^2$) × 50 psi (345 kPa)
500 × 50 = 25,000 lbf (111 kN)
$F_C$ = (40,000 – 25,000 lbf [178 – 111 kN]) 80% = 12,000 lbf (53 kN)

## Creasing (Mechanical Forging)

Creasing plastic by mechanical force in a thermoforming mold is used to create hinges in polystyrene foam (see Figure 7-13) and certain polypropylene parts (see Figure 7-14). The foam cross-section is thinned and made dense in the hinge area by squeezing its cross-section so it then becomes ductile and performs its hinge function without cracking. Direct thermoforming of polypropylene produces a standard hinge acceptable for most packaging functions. Mechanically forging the hinge creates a long-lasting product in which the hinges lay flat when opened.

## Venting

Fabricating vent holes for hot food containers can be as simple as fixing a sharp needle into the mold to pierce the plastic as it is

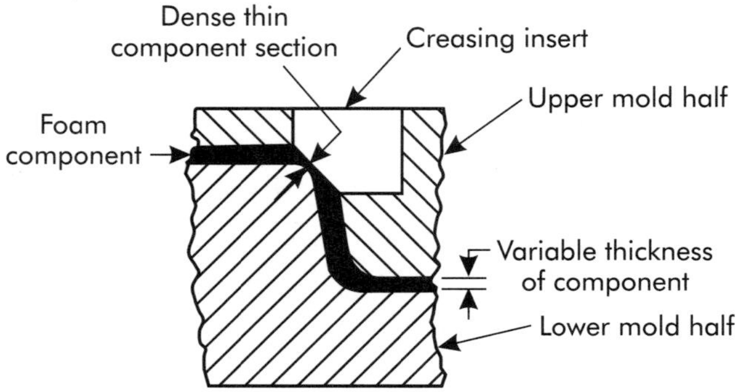


*Figure 7-13. Polystyrene foam crease for hinge.*

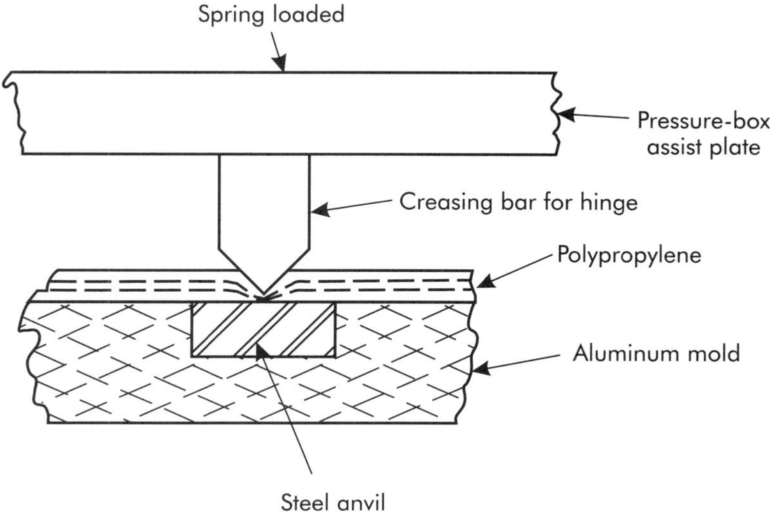


*Figure 7-14. Forging a living hinge in polypropylene.*

thermoformed. More elaborate vent designs may require shearing an opening using inserts built into each half of the mold. Either method ensures no plastic material breaks into loose fragments, which can accumulate on the mold and contaminate an incoming plastic sheet.

## Drinking Straws

Drinking straw openings and tear tabs for sipping beverages out of lidded cups are partially cut through the plastic. These partially trimmed apertures are produced while thermoforming by a knife-like die striking a hardened anvil in the opposite half of the mold. It is important to control the force of the punch striking the anvil to prevent rapid dulling of the knife and to assure the depth of trim is sufficient to open the aperture easily.

It may not be practical to make-ready (shim) each anvil in a multi-cavity mold, so each anvil may be loaded either with Belleville spring washers or a nitrogen gas spring to provide suitable resistive force.

## Screw-type Lids

Bottle-cap unthreading of screw-type enclosures can be built into multi-cavity thermoforming molds. The cavity is set in a bushing and a gear is fixed to each cavity. All the cavities in a row engage a rack driven by a cylinder or electrical stepping motor, which rotates the mold a fixed number of degrees. The driver is synchronized with the forming cycle so the cavity unscrews before the shot is ejected from the mold.

## Quick-change Tooling

Quick-change tooling-accessory devices speed tooling setup (see Figure 7-15). A thermoforming-press platen plate is machined to contain prelocated bushings. These bushings can be unlocked or latched to receive the self-aligning, locking knobs located on the pressure box and mold base. This system allows for tool preparation to be completed in an area away from the thermoforming machine, affording the machine more productive running time. The system predetermines accurate alignment of both halves of the tool, shortening setup time. Water-cooling, forming pressure, and vacuum supply hoses are quickly and easily attached to the mold if large-size, quick-disconnect fittings are used. Oversize quick-disconnect fittings are recommended because they do not reduce air or liquid flow. These measures can reduce mold changeover time from half a day to less than one hour in some cases.

## Estimating Initial Sheet Thickness for a Mold

The specified minimum wall thickness of the finished part affects the starting thickness of the sheet and its cost—a major element in the price of the part. One approach to estimating the starting sheet thickness is:

$$t = \frac{t_t \times A_c}{A_p} \qquad (7\text{-}2)$$

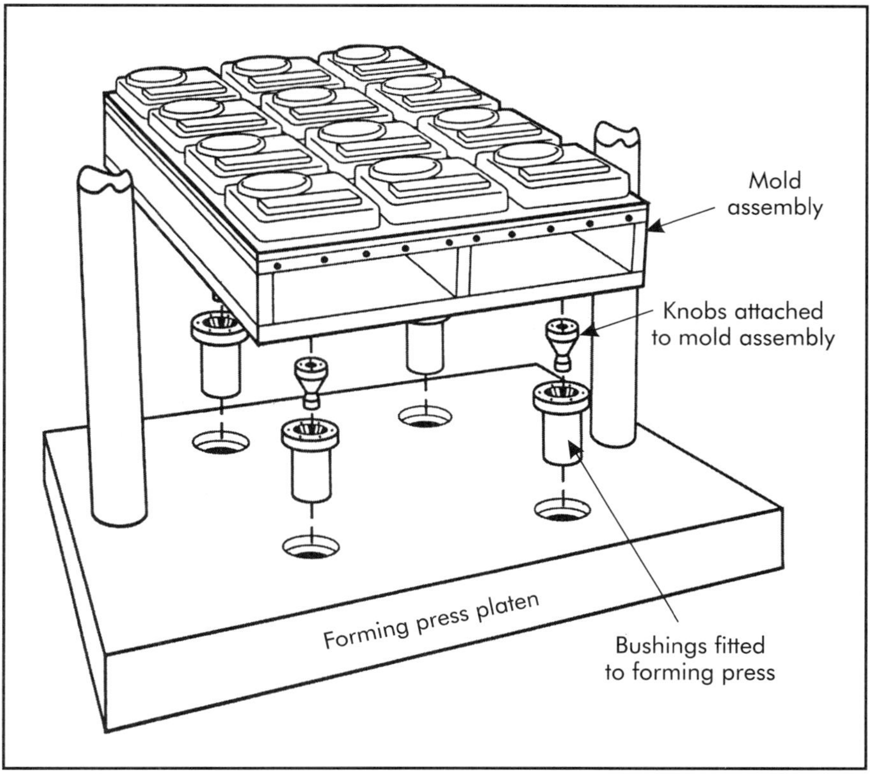


*Figure 7-15. Quick-change tooling accessories. (Courtesy E.D. Segen & Co., LLC)*

where:

$t$ = estimated sheet thickness required for thermoforming, in. (mm)

$t_t$ = theoretical desired thermoformed wall thickness, in. (mm)

$A_c$ = all of the cavity sidewalls' area + base area, in.$^2$ (m$^2$)

$A_p$ = heated projected sheet area available to cavity, in.$^2$ (m$^2$)

Estimate a given sheet thickness from the specified minimum wall thickness as if it were possible to thermoform a uniform wall thickness.

Wall thickness could be calculated assuming an even distribution covering the walls and base of a mold and by its projected sheet area above the cavity. For example, what is the starting sheet thickness of a rectangular cavity, which is 2 in. wide × 2 in. long (50.8 × 50.8 mm) and 1.5 in. (38.1 mm) deep, excluding radius and draft angles, and with a theoretical thermoformed wall thickness of 0.007 in. (0.18 mm)?

$A_p$ = 3 × 3 in. (76.2 × 76.2 mm) center-to-center between cavities = 9 in.$^2$ (0.006 m$^2$)

$A_c$ = (2 × 1.5)(4 sides) + (2 × 2)(bottom) = 16 in.$^2$ (0.01 m$^2$)

$$t = \frac{0.007 \times 16}{9} = 0.012 \text{ in. } (0.31 \text{ mm}) \text{ sheet thickness}$$

This calculation can be helpful in estimating the maximum and minimum expected wall thickness of the part. The maximum wall thickness appears on the flange of a female mold or at the top portion of the male cavity and is approximately starting-sheet thickness. The minimum thickness, depending on the efficiency of the plug or assist, may be 25–40% of the theoretical wall thickness. When calculated for the above example, the minimum wall thickness is 0.003–0.005 in. (0.08–0.13 mm). These differences in minimum wall thickness can be narrowed by designing efficient plugs or assists. For the male mold, increasing the mold center-to-center distance reduces starting sheet thickness ($t$) by increasing the heated projected area ($A_p$). Female molds generally utilize only the plastic sheet, which is directly above the cavity opening. Increasing or decreasing the center-to-center distance may have little effect on the wall thickness within a female mold.

## Conclusion

Mold design innovation is driven by industry demands for more complex product structures and the need to simplify the steps needed to produce the final product. When the designer visualizes a new tooling concept and puts the thought on paper (or CAD program), the resulting idea should be tested physically on a prototype mold. When the sample mold is demonstrated, the designer should carefully study every detail of the trial and not act on secondhand descriptions. There is much to be learned from the failure as well as the success of new procedures, and modification of an initial concept is more the rule than the exception.

# 8

# Mold Layout and Mold Base Design

This chapter will discuss mold layout/design, part shrinkage, pattern shrink allowance, male and female mold layout, venting of cavities, mold buildup layout, mold base design, and contact heat mold layout. Mold design can begin after the information discussed in Chapters 6 and 7 has been provided to the designer.

## Thermoformed-part Shrinkage

Many factors influence the sizing of a mold to correct for thermoformed-part shrinkage. These conditions, which the mold maker does not control, occur during thermoforming. They include the mold shrinkage coefficients, the type of resin used, maintaining a stable mold temperature, and any differential shrinkage in the direction of extrusion versus the width of the web.

### Mold Shrinkage Coefficients

Most roll-fed thermoforming machines are operated at high speed so that formed parts do not cool completely during their residence time in the mold. Some plastic resins may require 24 hours to shrink to a stable size.

Molds that do not maintain a stable temperature can change the thermal shrinkage rate by a factor of up to 20% when certain resins are thermoformed.

Sheet orientation during the extrusion process often produces a sheet with greater shrinkage in the direction of extrusion than across the web width due to roll windup tension. The shrinkage coefficients listed in Table 8-1 are a median figure, which can be used as a starting point for most molds. Very accurate sizing of a mold may require prototyping parts on a sample-making thermoformer using the same extruded stock as the production run.

**Table 8-1. Plastic shrinkage coefficient**

| Plastic Resins Unfilled | Mold Length Shrinkage Coefficient (*S*)* in./in. (mm/mm) |
|---|---|
| Polyvinyl chloride (PVC) | 0.004–0.005 |
| High-impact styrene | 0.005–0.006 |
| Polyethylene terephthalate glycol (PETG) | 0.007 |
| Polypropylene | 0.017 |
| High-density polyethylene | 0.022 in direction of extrusion 0.017 across web |

* Same coefficient can be used for either inch or metric system

For example: a cavity 10 in. (254 mm) long is to be increased in length to provide allowance for high-impact styrene shrinkage:

$$C_L = P_L + (P_L \times S) \tag{8-1}$$

where:

$C_L$ = cavity length, in. (mm)
$P_L$ = part length, in. (mm)
$S$ = shrinkage coefficient (from Table 8-1)
$C_L = 10\text{ in.} + (10 \times 0.006\text{ in.})$ or $254\text{ mm} + (254\text{ mm} \times 0.006\text{ mm})$
$C_L = 10.06\text{ in.}\ (255.5\text{ mm})$

## Pattern Shrinkage Allowance for an Aluminum Casting

The casting process requires making an impression of the pattern in sand at the foundry. A professional pattern maker builds the pattern larger by the amount of combined shrinkage that occurs when molten aluminum solidifies and thermoformed plastic cools.

When the cavity pattern is fabricated, the part drawing dimensions should include these shrink factors. For example, a high-impact styrene part pattern to be cast in aluminum has a shrinkage factor as follows:

Aluminum shrinkage = 0.013 in./in. (0.013 mm/mm)

Styrene shrinkage coefficient = 0.006 (from Table 8-1)

Combined shrinkage = 0.013 + 0.006 = 0.019 in./in. (0.019 mm/mm)

A 10-in. (254-mm) finished-plastic dimension on the pattern will require adding the shrinkage factor to the pattern dimension:

10 in. + (10 × 0.019) = 10.19 in. or 254 mm + (254 × 0.019) = 258.8 mm)

## Male Mold Layout

After deciding to build a male mold to produce a thermoformed component on a specific forming machine, the number of cavities can be computed. The machine model chosen determines the maximum mold area and allowable part height.

Some of the details that govern the center-to-center distance of the cavities are analyzed as follows.

Studying the geometry of the male cavity is a required preliminary for the layout process. Factors to be considered are the draft angle and maximum height of the cavity, as well as whether it has a symmetrical footprint.

Calculating the center-to-center cavity dimensions provides information related to the maximum number of male cavities. Figure 8-1 shows where the data is obtained and Equation 8-2 shows how to calculate the layout dimensions.

$$D = A + B$$

$$B = [L - 2(F \times \tan E)]$$

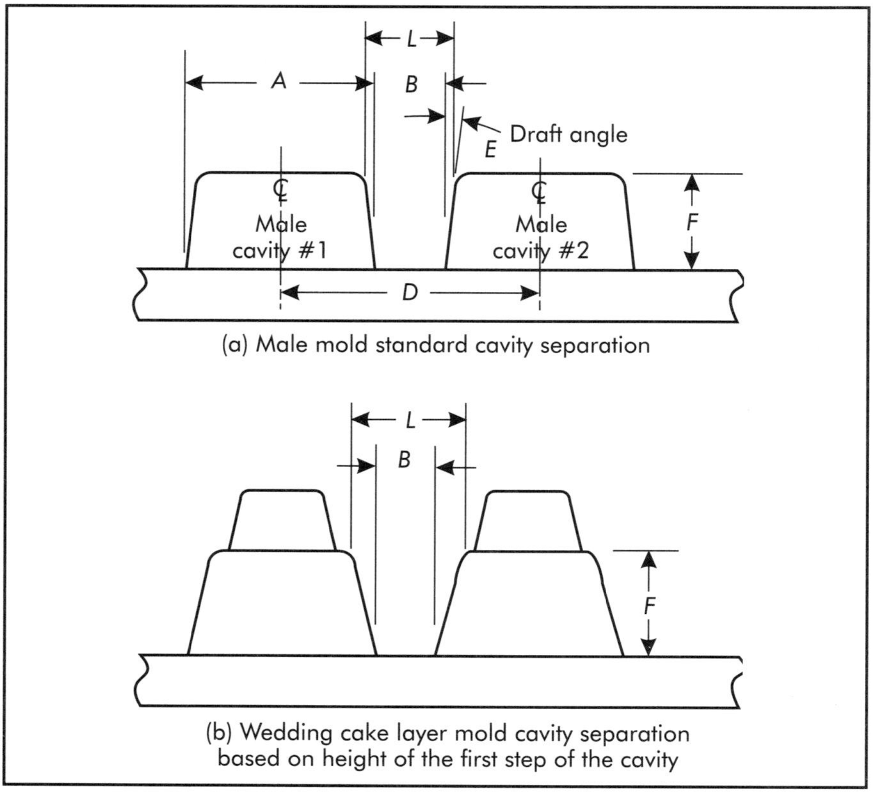


*Figure 8-1. Calculating the center-to-center distance between male cavities' D dimension (Mitchell 1998).*

$$D = A + [L - 2\,(F \times \tan E)] \qquad (8\text{-}2)$$

where:

$D$ = center-to-center of cavities ($A + B$), in. (mm)
$A$ = base dimension of the cavity, in. (mm)
$B$ = separation of cavities measured at the base of cavity, in. (mm)
$L$ = $F$ for small draft angles less than 5°
or
$L$ = $F \times 0.75$ for draft angles greater than 5° (maximum angle used is 10° for this calculation)
$F$ = cavity height, in. (mm)
$E$ = draft angle of cavity, °

For example:

$A$ = 3 in. (76.2 mm)
$E$ = 5°
$F$ = 2 in. (50.8 mm)
$L$ = 0.75$F$ (when $E$ = 5° or greater) = 0.75 × 2 = 1.5 in. (38 mm)
$D$ = $A$ + [$L$ – 2($F$ × tan $E$)]
= 3 + [0.75 (2) –2(2 × 0.088)] = 4.15 in. (105.4 mm) center-to-center of cavities

Other considerations for center-to-center location of cavities may include:

- Symmetrical cavities, including round parts, are best placed in straight in-line rows as shown in Figure 8-2. Although the most economical sheet layout for round parts is an equilateral triangle arrangement, thermoforming presses and index devices are not suited for handling this type of layout.
- Polygons should be nested to make the best use of the mold area (see Figure 8-3).
- Alternating high- and low-profile mold sections helps to distribute the wall thickness more uniformly among the cavities (see Figure 8-4).

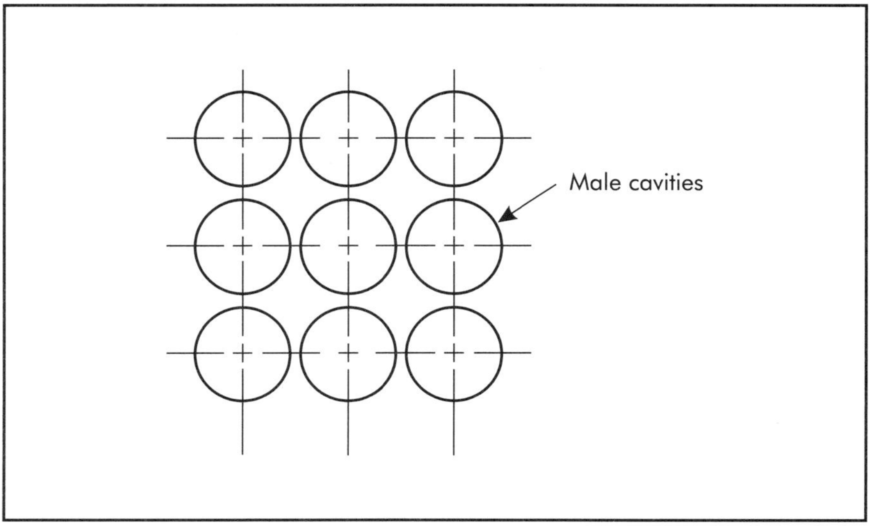


*Figure 8-2. Symmetrical male cavity layout.*

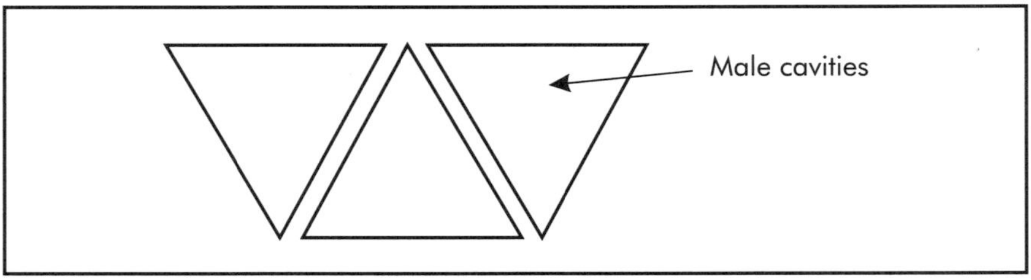


*Figure 8-3. Polygon male cavity layout.*

- The specified part flange width required for the finished, trimmed components may be greater than the calculated value of $B/2$ (see Figure 8-1). The flange width specifications would then control the layout rather than the dimension $B/2$.

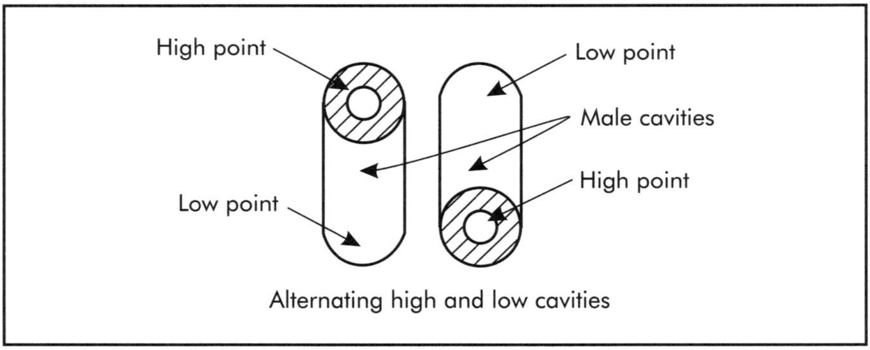


*Figure 8-4. Alternating high- and low-profile male cavity sections (Mitchell 1998).*

## Trim Die Mold Clearance

Trim-die minimum cut-to-cut physical requirements should be considered at this point in the design layout study.

### *Steel Rule Die Allowance*

Steel-rule-die knife blades are held in a high-quality laminated wooden die board by friction only and sufficient wood must be in place for the die to maintain its structural integrity. Die makers recommend 0.25 in. (6.4 mm) minimum of wood between blades for straight sections and somewhat less for round configurations. Some formed components can be located so that a single common knife serves to separate two adjacent parts. Under these conditions, there is no skeleton scrap web to help transport trimmed parts out of the machine for packing. Parts with a common cut are attached by narrow bridges (nicks) of uncut plastic at several points in such a manner that the connection between the attached parts is strong enough to transport, yet weak enough to separate at the packing station.

## *Forged Die Allowance*

Forged high dies are individual knife-like dies with wall thicknesses of 0.12–0.25 in. (3.0–6.4 mm), depending on the height and die shape. The die walls can be modified to reduce their thickness for closer center-to-center cavity placement because two abutting die walls tend to support each other. A fairly safe cut-to-cut minimum dimension is 0.32 in. (8.1 mm).

## *Punch and Die Allowance*

Center-to-center minimum-die-wall widths for the punch and die depend on several factors. Round dies are weakest at their points of tangency so that small-diameter dies may require less separation, 0.19 in. (4.8 mm), and larger dies may require 0.50 in. (12.7 mm). The minimum width of straight-wall rectangular dies depends on the length of the wall and die thickness. Long, unsupported sidewalls deflect sideways under load more easily than shorter lengths, thus increasing die clearance and disturbing die-trimming action. A straight die wall of 6 in. (152.4 mm) length may require a wall width of 0.50 in. (12.7 mm). Longer die walls can require widths up to 0.75 in. (19.1 mm) or more to stiffen the cross-sectional area.

Multi-row punch and dies are used when thermoforming output would overwhelm a single-row die being trimmed at maximum speed on an off-line trim press. The layout in the index direction of cavity rows should be an even multiple of the number of trim rows so the die always trims product from a single shot. If a multi-row die straddles the last row of one shot and the front row of the incoming shot, the thermoformer index stroke must be perfect, or the trim cut will be out of registry. For example, eight rows in one shot requires four trim strokes trimming two die rows per press stroke, which is the correct layout. Nine rows in one shot requires 4-1/2 trim strokes trimming two die rows per press stroke, straddling two shots, which may cause registry error.

### *Indexing Pusher Lug Allowance*

Indexing pusher lugs are used to advance one-or-more formed rows into the cutting area of the punch and die by the trim press indexer (see Chapter 5, Figure 5-9). These lugs are placed away from the flange trim perimeter to avoid marking the finished product and located so as to decrease the web waste area. Lugs are often formed on the opposite face of the web from the thermoformed part and placed so they do not interfere with the web guide bars. Once the formed web leaves the thermoformer, it is free of the thermoformer chain index until it is picked up by the off-line trim-press indexing system. The formed sheet is guided into proper cross-web position by a series of guide bars that can be placed in contact with the sidewalls of the formed parts. Usually two or more of these guides are necessary to locate and support the web and space should be allotted for these guides, which are approximately 0.50 in. (12.7 mm) wide (see Figure 8-5).

## Fixed-Mold-Base Requirements

Fixed-mold-base space requirements are the same for narrow or wide molds. The most economical use of plastic is to design for maximum mold width and length (see Figure 8-6).

### Machine Chain Width and Index Length

The thermoforming machine manufacturer specifies the maximum inside chain width, which is the starting point for layout of

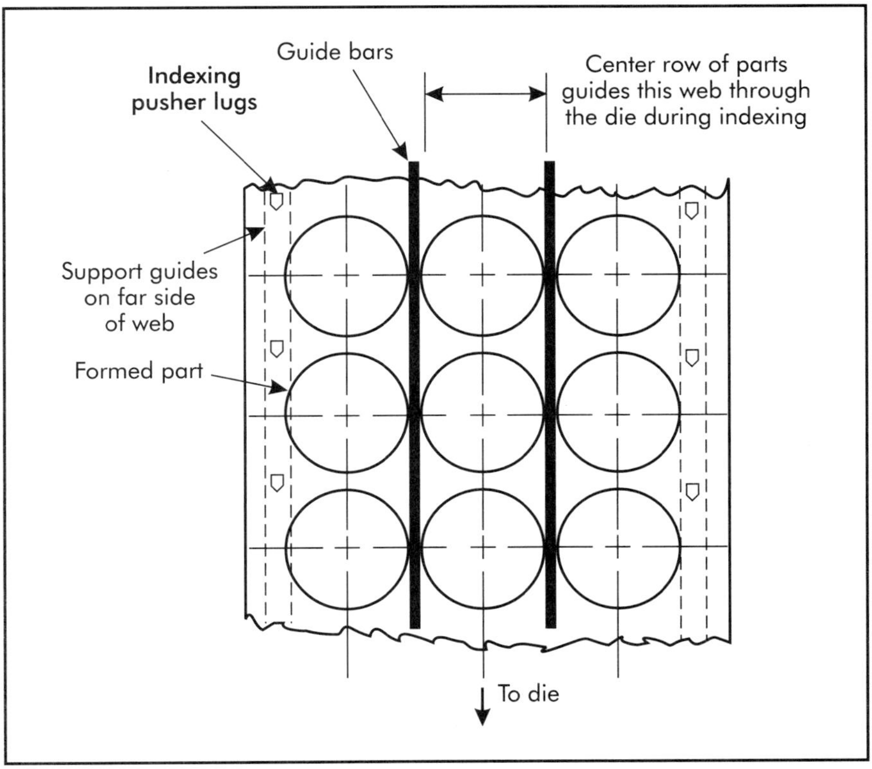


*Figure 8-5. Indexing pusher lugs register the web with respect to the die.*

the maximum-width mold. The minimum specified width dimension between chains can be surprisingly large, and a small-width mold may require an oversize mold base to fill available space. Uncooled, excess sheet-edge overhanging the mold can cause wrinkling of the web and other thermoforming problems if a smaller-than-recommended, minimum-mold-width base is used.

Thermoforming machine operators often attempt to reduce sheet sag by setting the chain rails to diverge from back to front so that the entrance chain-width dimension is smaller than the exit distance. This divergence can be up to 1 in. (25.4 mm) overall, so when designing the maximum-width mold, this factor should be considered. Minimum side clearance of 0.12 in. (3.0 mm) per sheet edge is recommended between the index chain and mold base to avoid rubbing contact.

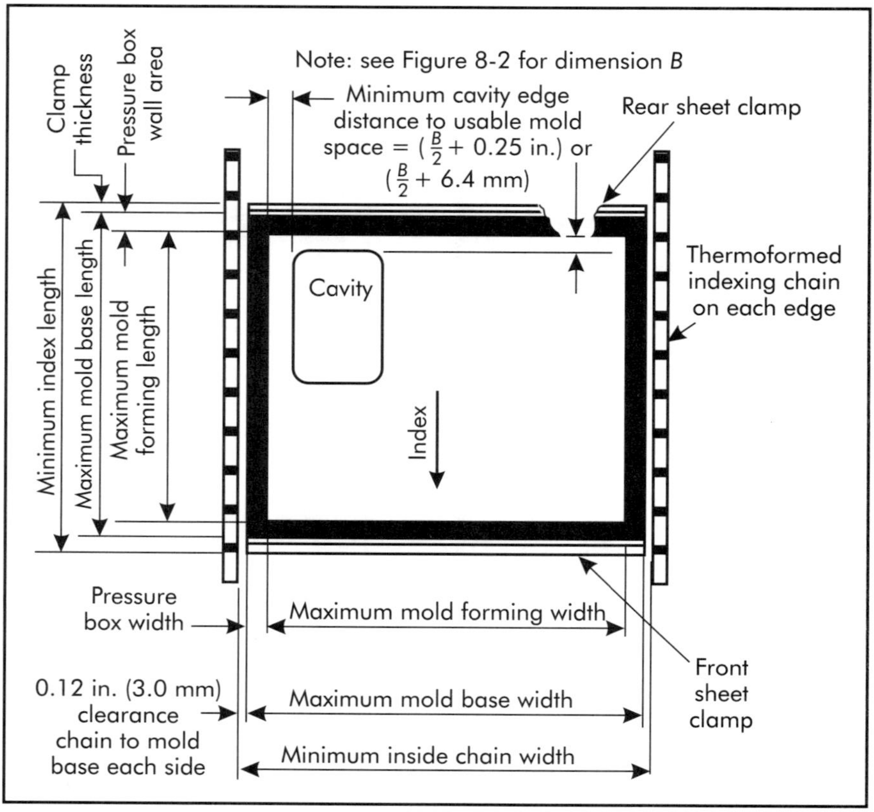


*Figure 8-6. Fixed-mold-base space requirements.*

To calculate the minimum web width after selecting the mold-width dimensions:

$$W_w = m_b + C_c \tag{8-3}$$

where:

$W_w$ = minimum web width, in. (mm)
$m_b$ = mold base width, in. (mm)
$C_c$ = chain side clearance and chain grip minimum = 0.875 in. per side (22.23 mm/side) = 1.75 in. (44.5 mm) total

For example:

$m_b$ = 25 in. (635 mm)

$C_c$ = 1.75 in. (44.5 mm)
$W_w$ = 25 + 1.75 = 26.75 in. (679.5 mm)

The sheet web may not track true on the chain, so the chain grip includes a ±0.25-in. (±6.4-mm) web tracking error. If the chain grip clearance is increased by a large amount, the sheet edge may become entangled in the chain assembly, causing an indexing stoppage of the thermoformer. Some thermoforming machines are not capable of properly controlling the tracking of the web through the machine and may require a broader edge allowance than 1.75 in. (44.5 mm).

The in-line thermoforming machine manufacturer specifies the web's maximum index length. This can become a very critical dimension. Forming press to trim press separation is often an exact increment of index length, and some machines do not have leeway beyond the maximum index length to accommodate this little bit extra, which is sometimes needed. The web index length can be set up so that the most economical use of the web would be for the rear clamp to index to the front clamp position on the mold base.

## Pressure Box

The pressure-box seal-off area on the mold often cannot be utilized as part of the thermoforming area because of distortion caused by pressure-box silicone rubber seals. Inside the seal-off area is the usable mold area—the space allotted for layout of the cavities.

The layout of the edge distance of the first cavities inside the usable mold space is the first step in mold design. A male mold should maintain a minimum allowance of $B/2$ + 0.25 in. (6.4 mm) distance from the cavity edge (see Figure 8-1 for "$B$" location) to inside the pressure-box inside wall (see Figure 8-6). A low-profile (less than 1.00-in. [25.4-mm]) male cavity that requires an assist needs a minimum of a 0.38 in. (9.7 mm) clearance from the pressure-box inside wall.

Male molds that are 2.50 in. (63.5 mm) or more in height may require a four-sided sheet clamp to prevent the hot web from being torn from the chain as the male mold enters the web. A four-sided clamp requires an additional minimum of 0.32 in. (8.1 mm) per chain side of space on the mold layout. Its open-box structure is comparatively weak because its sides are joined at the corners by very-small-diameter fasteners. An alternative method to avoid using the four-sided sheet clamp is to increase the cavity-edge distance sufficiently to provide additional hot-plastic web width, which acts like a rubber band, reducing the chain-stripping force.

Female cavities that require a plug and do not use the trapped-sheet method of thermoforming may need a four-sided sheet clamp attached to the pressure box. A plug can act in a manner similar to a tall male mold to pull the web free from the chain. If the female cavity is shallow or does not require a plug, it can be located as close as is practical to the mold edge.

## Female Mold Layout

A female mold cavity may be mounted on either the upper or lower moving platens of a thermoforming press, depending on conditions. Open-and-shut maximum travel of both platens may not be equal, so that a deep-formed component may clear the mold assembly on only one side of the sheet face.

It is usually advantageous to mount the female cavity on the top forming-press platen to satisfy most thermoforming functions. A hot plastic sheet sags due to the force of gravity and its surface area increases. The lower platen plugs take advantage of this effect and distribute available plastic uniformly to the female cavities (see Figure 8-7). In-line-thermoforming trim presses are often designed so that the deepest parts clear the upper trim-press platen, which facilitates locating female cavities on the top forming-press platen and male cavities on the lower platen.

Off-line trim-press dies can favor the finished thermoformed-part profile that is presented toward either a flat or female cavity punch for piloting or indexing (see Figures 8-8 and 8-9). The trim-press assembly can be manually shifted to accommodate this preferred direction. But this relocation can cause the human part packer to work between the thermoforming machine and the trim press. This position may be inconvenient and uncomfortable to the packer and make it difficult to freely access the flow of materials. A choice must be made among these competing factors to

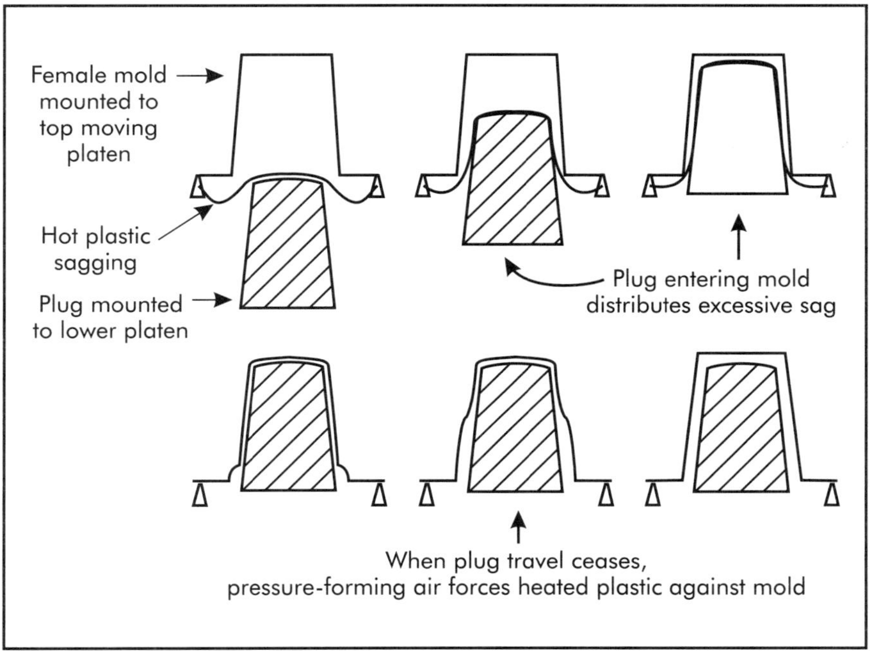


*Figure 8-7. Action of a plug entering a female mold distributes excess sheet sag (Mitchell 1998).*

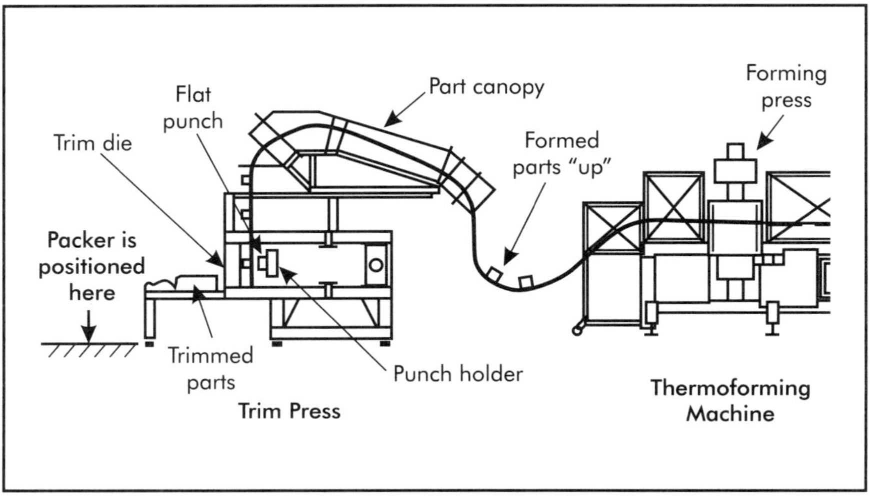


*Figure 8-8. When using a flat punch, the packer is clear of the machinery (Mitchell 1998).*

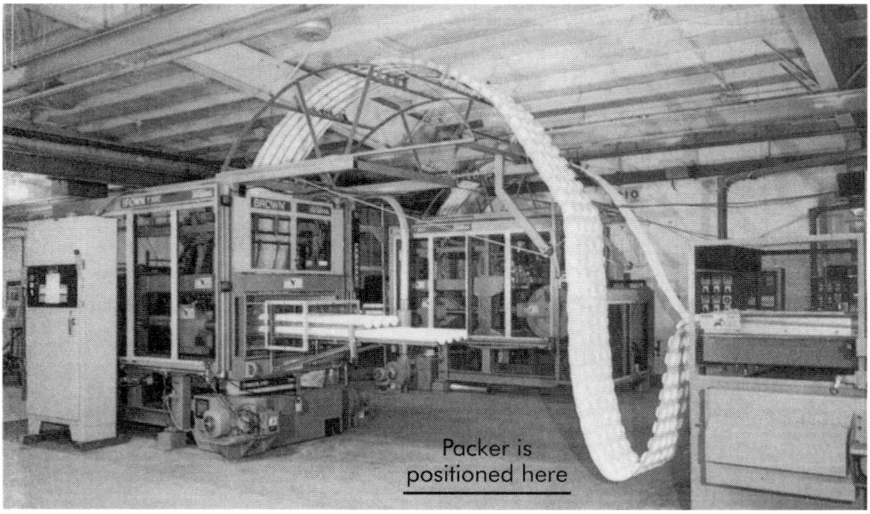


*Figure 8-9. When using a female piloted punch, the packer is positioned between the machinery. (Courtesy Brown Machine, LLC)*

prioritize the platen-mounting location of female cavities on the thermoformer.

## Arrangement

Female multi-cavity molds can be machined directly into a single aluminum mold block if this is the preferred fabricating choice. It is the most efficient mold heat-transfer method (see Figure 8-10).

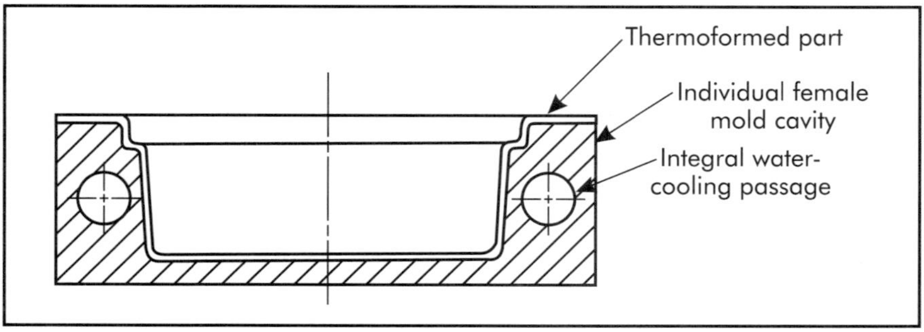


*Figure 8-10. Female cavity machined directly into a mold block with integral cooling.*

Complex multi-cavity shapes often are more economical to cast or machine as individual loose cavities. This is also the case with cavities later interchanged in a family mold.

Loose, individual female cavities with square or rectangular outer perimeters can be mounted in an open box-like multicavity mold, the walls of which align with the pressure box and clamps. A V-groove is machined into the front and rear walls to receive the sheet clamps. The cavity vent space of 0.015 in. (0.38 mm) should be maintained around each cavity perimeter to allow entrapped air between cavities to be vented (see Figure 8-11). Cooling relatively shallow cavities (2 in. [50.8 mm] or less) can be accomplished by mounting the box containing the cavities on a temperature-controlled mold base in the same manner as a male mold. Deeper (greater than 2 in. [50.8 mm]) molds may require individual water-cooling circuits to each cavity.

Deep round-perimeter cavities can be placed in holes bored into solid aluminum cavity mold blocks, sealed with O-rings at the open ends, and flood cooled by water passages drilled into the block (see Figure 8-12).

## *Undercuts*

Cavities with undercuts are most often formed in female molds because the plastic cools and shrinks away from the cavity walls, assisting the release of product from the undercut mold segment.

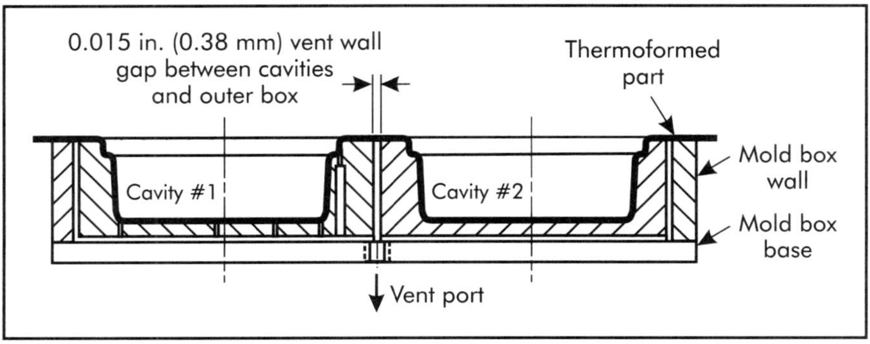


*Figure 8-11. Individual female cavities mounted in a mold box with a vent gap around the perimeter.*

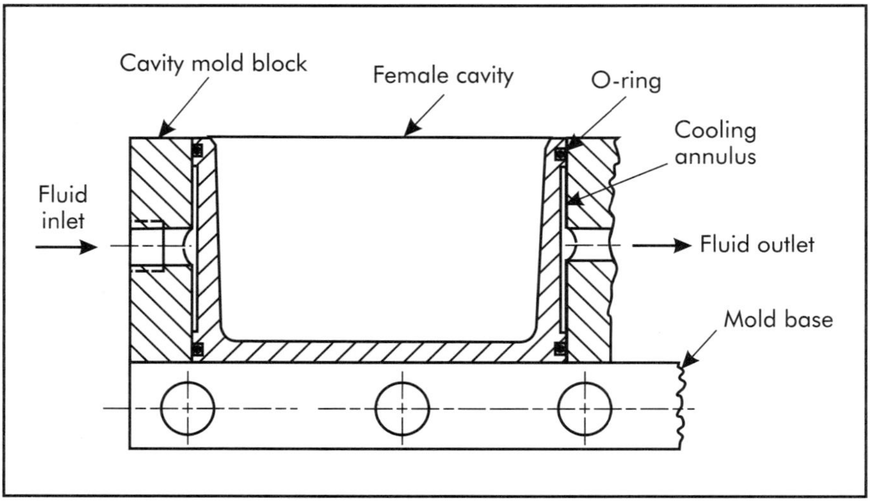


*Figure 8-12. Cooling annulus for a deep female mold.*

If the undercut section is more than 0.03 in. (0.8 mm) deep and is continuous around the cavity, it may prove difficult to strip the part undamaged without mechanical assistance. This determination should be made during the sampling stage because it is too complex to improvise a knockout on a completed production mold. If the undercut can be interrupted for 50% of its length, it may prove much easier to eject from the mold.

If the undercut is located near the upper portion of a female cavity, it is best to split the mold so that separation of the top half (knockout plate) is just below the undercut (see Chapter 6, Figure 6-16). A knockout plate, activated by air cylinders or a mechanical mechanism, should eject the part at the same instant the platen retracts away from the sheet line. The knockout plate may require water cooling as it is in direct contact with the hot sheet for long periods of the forming cycle. Cavity spacing center-to-center should take into consideration the area needed for water-cooling passages in the knockout plate.

A knockout pin through the base of the mold may more easily eject undercuts located near the base of a female cavity (see Chapter 6, Figure 6-15). The larger the head area of the knockout, the greater is its effective stripping action. The head also can be designed as an integral part of the mold aesthetics. If the head is relatively small compared to the force needed, it may punch through the plastic wall or wrinkle or distort the finished component. The knockout pin is cooled by contact with the main mold body.

## Venting of Cavities During Thermoforming

Both pressure and vacuum forming have the same design criteria to provide exits for air trapped between hot plastic and cavity walls. The greatest volume of air is initially evacuated as the hot sheet stretches and advances rapidly toward the walls, displacing the air in the cavity. This initial action requires sufficient cavity vent area, consisting of small holes or slots and a system of air channels, to funnel air toward the exhaust valve port. Finally, the expanded sheet may entrap pockets of air against the cavity wall, which then can be vented to an exit by escaping through hills and valleys created by a roughened mold-surface finish.

If the plastic freezes and can no longer easily stretch due to slow evacuation of entrapped air, the resulting formed part may not correctly conform to the detailed contours of the cavity. A female cavity must have sufficient vent area to rapidly exhaust air from the cavity when a plug mechanically stretches the sheet just before vacuum or pressure forming commences. If internal cavity air pressure is not quickly vented, the advancing plug increases pressure between the hot sheet and cavity wall. Pressure buildup may burst the sheet, thereby ruining the shot. A method of decreasing cavity air pressure as the plug enters the cavity is to introduce a partial vacuum of 5–8 in. Hg (127–203.2 mm Hg) of mercury (approximately 0.25 atmosphere) to help exhaust excess pressure. This technique requires a machine operator with considerable experience and skill. If the vacuum is applied too soon or with excessive strength, the sheet will be thermoformed by the vacuum, losing the distributive advantage of the plug. The resulting part may be formed with very thin lower walls and bottom surface.

### Venting Methods

Techniques to allow air to escape from the mold include the use of vent holes, slot vents, or channel vents.

Drilling vent holes always leaves telltale marks at their openings on the mold side of the formed plastic, with the smallest drill size producing the least noticeable blemish. The minimum practical drill size is a #76 (0.02 in. [0.5 mm] diameter), which in unskilled hands can be very difficult to master using a standard drill press. A typical vent hole of 0.02 in. (0.5 mm) diameter is drilled three diameters deep (0.06 in. [1.5 mm]) and a mating, but larger, back-drilled hole completes the vent (see Figure 8-13).

## *Vent Holes*

Vent holes up to 0.06 in. (1.5 mm) diameter can be used on thicker opaque plastics hidden in among textured mold finishes. Care must be exercised in selecting a drill diameter for a specific resin (see Table 8-2). Some resins become so weakened when heated that they form icicles when pressure formed over larger-diameter vent holes.

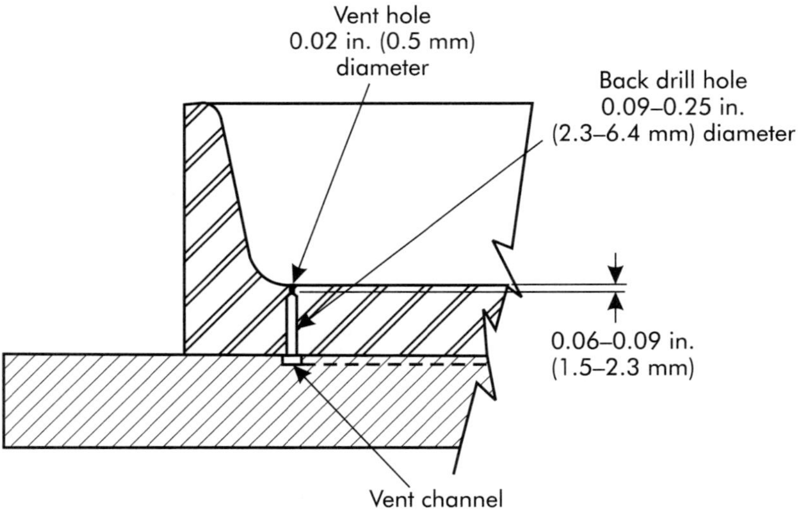


*Figure 8-13. Back drill to vent hole. (Courtesy G.N. Plastics Co., Ltd.)*

Table 8-2. Recommended drill sizes for venting

| Suggested Drill Size | Plastic Resin |
|---|---|
| Smallest diameter<br>0.02 in. (0.5 mm) | High-density polyethylene (HDPE)<br>Polypropylene (PP)<br>Polyethylene terephthalate (PET)<br>Clear polyvinyl chloride (PVC)<br>Oriented polystyrene (OPS) |
| Medium diameter<br>0.02–0.035 in.<br>(0.50–0.89 mm) | High-impact polystyrene (HIPS)<br>Acrylonitrile butadiene-styrene (ABS) |
| Largest diameter<br>0.035–0.06 in.<br>(0.89–1.5 mm) | Styrene and polyethylene foams and flocked high-impact polystyrene |

Note: A 0.06-in. (1.5-mm) diameter hole has an area equal to nine 0.02-in. (0.5-mm) diameter vent holes, which suggests using the maximum drill size appropriate for the cavity.

When venting engraved portions of the cavity, raised characters are self-venting; but sunken (embossed) sections may require a small hole in each letter for crisp reproduction.

## *Slot Vents*

Slot vents are the most effective method of exhausting entrapped air from both male and female cavities. The male-cavity slot vent is located at the flange junction to the formed part and is relatively simple to fabricate. Larger-diameter exhaust holes (0.19 in. [4.8 mm] diameter) are hidden under the base of the cavity, resulting in an almost invisible vent line in the finished part (see Figure 8-14). The same type of venting technique can be replicated at other points along the male or female cavity wall wherever the mold can be split (see Figure 8-15).

Cavity inserts, or other changes in a flat plane, offer an excellent location for slot venting both male and female molds. These locations offer a break point to disguise the slot so that it is not noticeable on the formed component (see Figure 8-16).

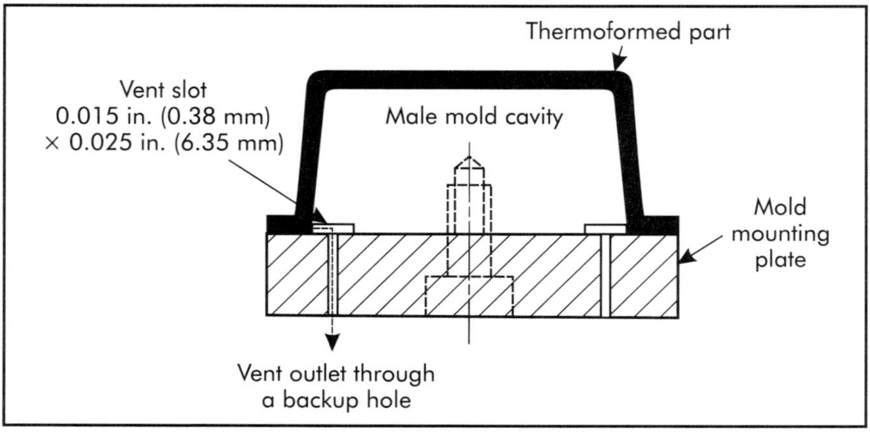


*Figure 8-14. Male cavity base vent slot is exhausted through a large-diameter backup hole (Mitchell 1998).*

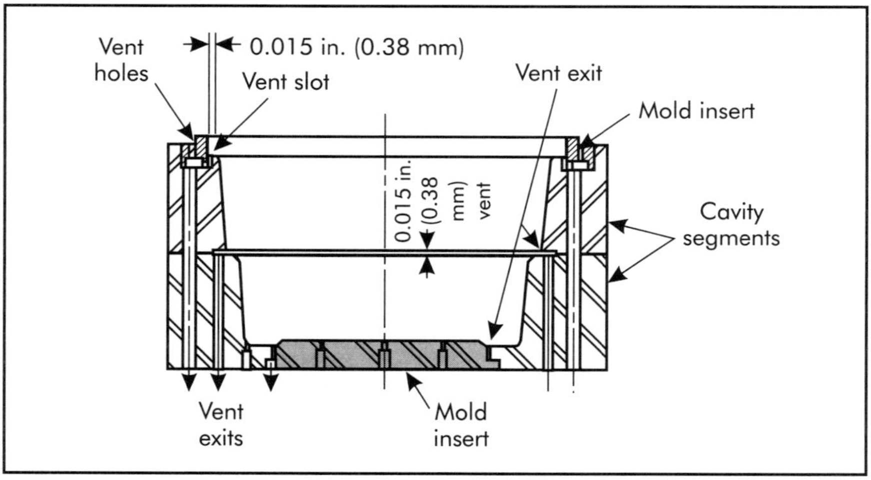


*Figure 8-15. Various methods of venting female cavities.*

## Channel Vents

Vent holes and slots direct exhaust air to channels cut into the back face of each cavity and toward a central collection spot (see Figure 8-17). The backside of the cavity is a conduit for heat to the mold base or mold plate so that channels and collection spot face-areas deduct from the available heat-transfer surface. These

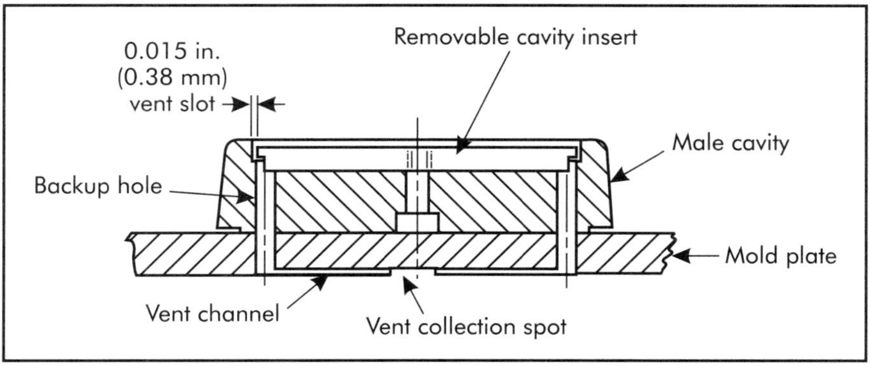


*Figure 8-16. Venting removable inserts in a male cavity plate mounted to a mold.*

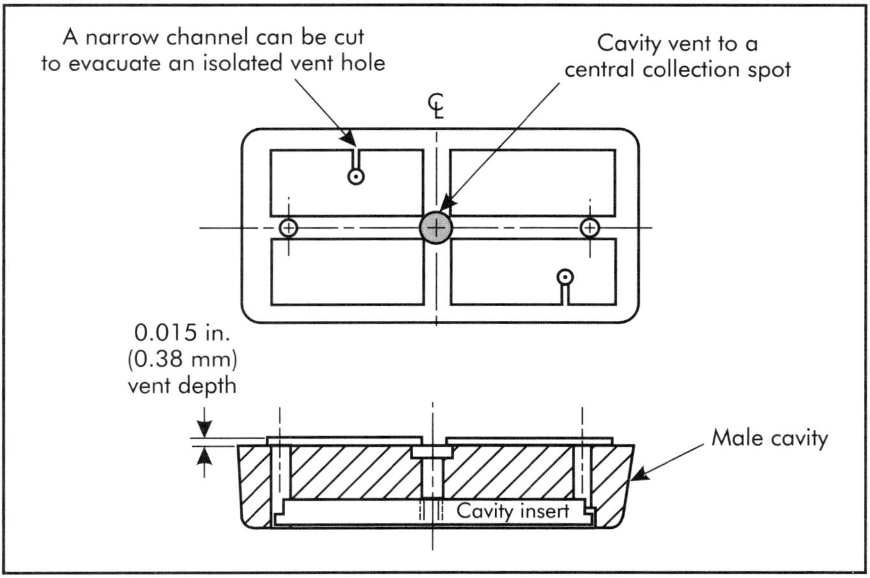


*Figure 8-17. Vent channels lead to a central collection spot on a male cavity.*

air-venting highways are best designed deep and narrow (except at the cavity perimeter) and take up a minimum of contact area to maintain maximum heat-transfer surface area.

When cavities are to be mounted on a mold plate, a cavity collection spot face is located over a drilled-through hole that terminates on the back face of the mold plate (see Figure 8-16). These holes are connected by channels in the back face of the mold plate to a central collection spot face that empties all the cavities into the mold-base vacuum port.

Male cavities are fastened directly to a mold base and vented through the mold base to a vacuum seal-off plate attached to the back face of the mold base (see Figure 8-18). Direct attachment of a cavity to a mold base offers the best heat-transfer method for removing heat from a male cavity less than 2 in. (50.8 mm) high. Family-mold members with the same footprint can share the mold base without any further modification.

Female cavities less than 2 in. (50.8 mm) deep can be cooled quite efficiently and vented easily when mounted directly to a mold base and enclosed by a perimeter wall. This layout allows re-use of the mold base for any female cavity that can be fitted within the walls by temporarily sealing off previous cavity-mounting holes.

Both male and female cavities deeper than 2 in. (50.8 mm) may require individual internal fluid circulation passages to achieve economical cycling rates (see Figure 8-19). This type of mold base is designed with two separate fluid temperature-control circuits: one to supply individual cavity circuits, the other to conduct heat from the mold base. O-ring seals around the inlet and outlet fluid ports prevent leakage when fastened closely with a minimum of three fasteners.

An arbitrary break point of 2 in. (50.8 mm) of cavity depth or height can be used to determine whether to cool by means of internal cooling passages based on some of the following considerations:

- A cavity can be cycled only as rapidly as the plastic can be cooled to the point where the shot can be ejected without damage.
- A high-profile part takes longer to transfer heat from the top face of the cavity to the mold base. Much of the hot-plastic-sheet thickness is deposited at the top face of the cavity, causing the greatest heat input to travel the furthest distance to be cooled. Thicker sheets (heavier weight of hot plastic to be cooled) and fast cycling may cause the top of the shot to be overheated unless a secondary cooling system is supplied close to the upper face of the cavity.

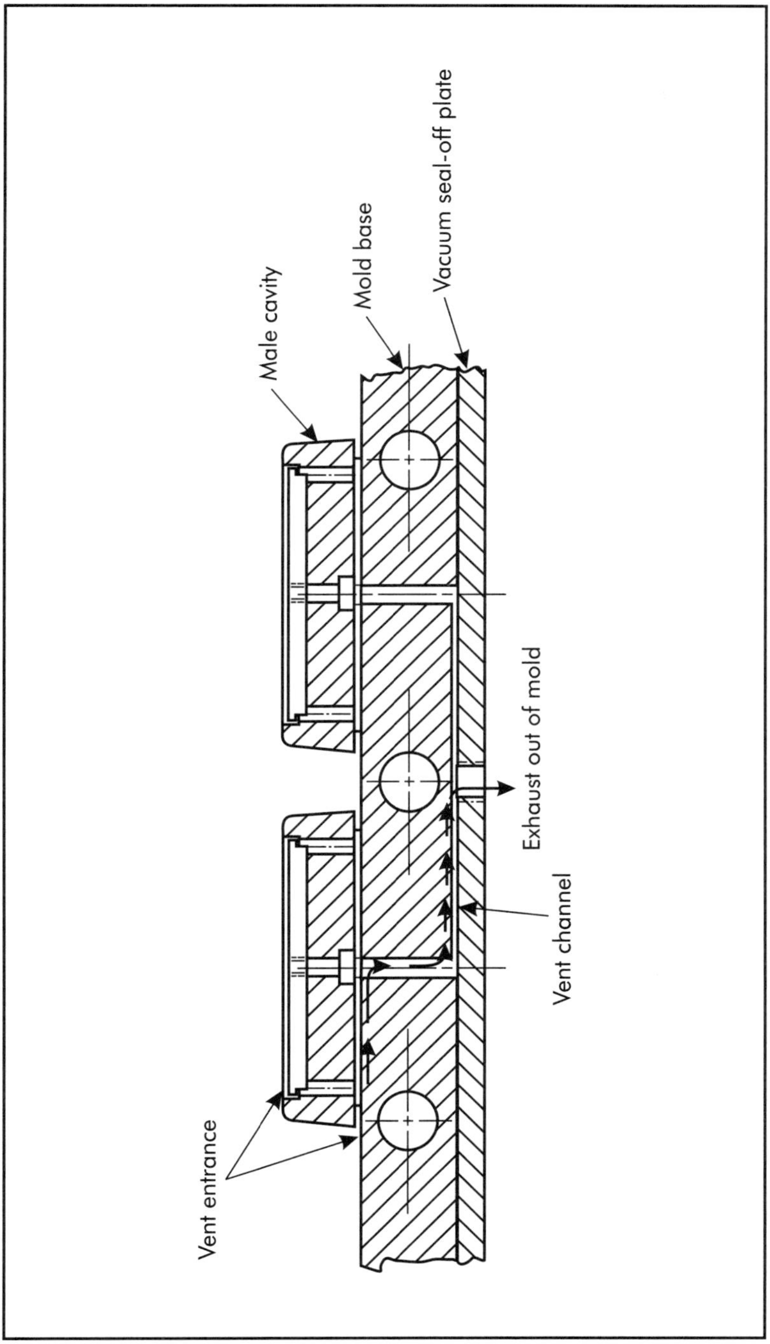


*Figure 8-18. Male cavities mounted directly to the mold base.*

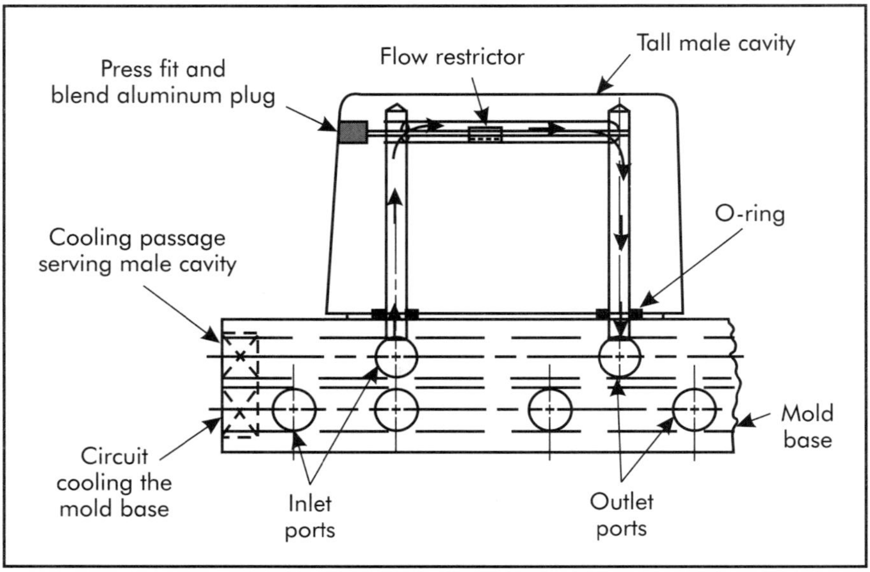


*Figure 8-19. Male cavity internal fluid circulation passages and mold base circuit.*

- It may be more expensive to provide internal cavity cooling than it is to slow down the thermoforming cycle.

## Cavity Fastener

The mounting bolts, which fasten the cavity to the mold base or mold plate, should be properly spaced and torqued to prevent air spaces from occurring between metal faces. Air acts as a heat insulator and prevents efficient metal-to-metal heat transfer. High-quality aluminum plates (jig or tooling) of precision thickness and flatness should be specified, and all cavity back faces should be machined flat for optimum heat-transfer characteristics. Aluminum and epoxy cavities, as well as mold bases and plates, are relatively soft materials, and, when tapped for small diameter fasteners, the threads can be easily stripped. Steel self-tapping threaded inserts are highly recommended for all bolts smaller than 0.375 in. (9.53 mm) diameter to prevent cross-threading by inexperienced thermoforming machine operators.

## Mold Buildup Layout

The machinery manufacturer provides envelope dimensions to which the mold base and pressure box (or vacuum seal-off box) are mounted to the press platens.

### Forming-press Specifications

Details considered for a forming press include open and closed press specifications, press toggle lockup, depth of draw specifications, maximum allowable tool weight and location, and mold clearance from the oven.

Platen fully retracted (press open) minus platen fully extended (press closed) equals the platen stroke. The dimensions for the shut height of the press are usually specified with a plus tolerance, minus zero, so that the actual shut height may require adjustment for proper mold lockup. Many forming presses possess a mechanism to adjust the upper platen shut height sufficiently to lock the tooling. Older presses may require shimming the tooling to close the open gap. Rather than having thermoforming machine setup personnel hunt for odd bits of shim stock, it expedites tool installation to provide various thicknesses of sheet metal with the platen bolt-hole pattern punched out for easy use.

#### *Press Toggle Lockup*

Most forming presses produce their lockup force by toggle action. Toggles deliver maximum power when both links are nearly fully extended perpendicular to the platen as per machine specifications. The toggle loses lockup power rapidly when it is not fully extended, so maximum tooling height should be kept within press lockup specification to prevent venting the air in the pressure box.

#### *Depth of Draw*

Some machine manufacturers may specify a maximum depth of draw, which describes the recommended maximum height of the part to be formed. Depth of draw and platen stroke can be different, so the lesser of the two numbers can be used for the maximum allowable part height.

### *Tool Weight*

Mold weight per platen can be estimated as soon as a preliminary tooling layout is available. It can be checked against the machine's specified maximum for each platen to prevent the press from being overloaded.

### *Tooling Location*

Locating tooling central to the press-platen centerline is important when pressure forming to avoid unbalanced opening forces, which could cause air leakage. However, forming presses, which are of four-post design and have a platen length greater than the mold length, may not accommodate shorter-length molds central to the platen. Machine builders claim the four-post design eliminates the problems of operating smaller pressure boxes off-center because the presses are capable of handling these heavy eccentric loads.

### *Oven Location*

The edge of the mold closest to the oven exit requires only 0.12 in (3.0 mm) clearance; the mold base should extend to overhang the rear edge of the platens. If the molds are located inside the press platen edge, the sheet area between the oven exit and mold back wall is cooled by air. When the shot is next indexed into the mold, that section of the sheet may not be hot enough to thermoform properly (see Figure 8-20).

## Mold-base Design

The mold-base design can begin once the mold layout has been completed and its overall dimensions are fixed. The major function

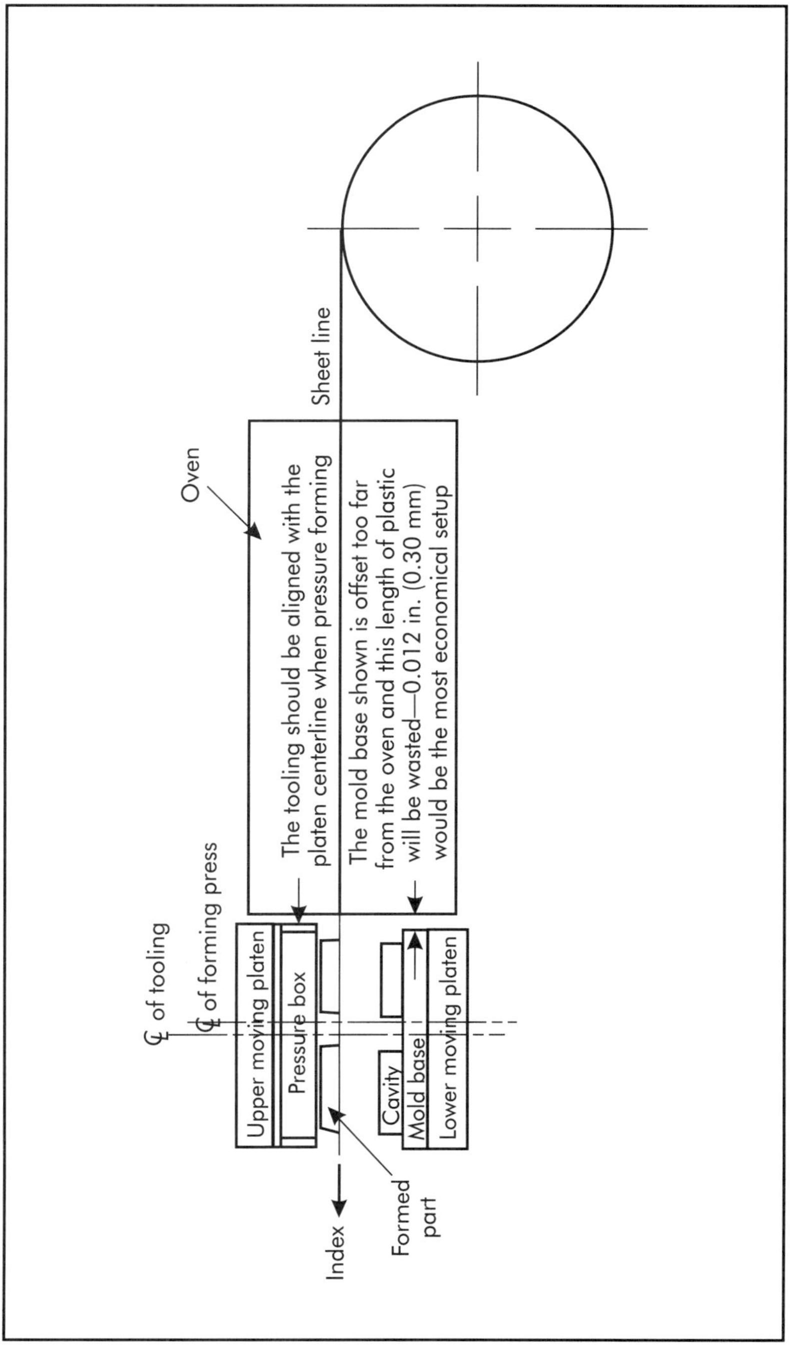


*Figure 8-20. Location of the back of the mold base relative to the oven.*

of the base design is for heat transfer from the cavities to fluids circulating through the mold base. It also serves as the structural foundation to which ancillary components such as cavities, sheet clamps, spacers, mold plates, and knockouts are attached.

## Thickness

The minimum thickness of a mold base is determined by the hole diameter required for its fluid passages, plus enough mass to dissipate the heat load of each shot. Two independent fluid circuits may be needed when fluid is both directed internally up into the individual cavities and used for cooling the mold base (see Figure 8-19). Pneumatic or vacuum actuators may be built into the mold base or cavities for articulating mold parts and to power knockouts. A separate, drilled, integral compressed-air circuit can supply the actuators (see Figure 8-21). Mold-base temperatures over 250° F (121° C) are required to thermoform polysulfone or crystalline polyethylene terephthalate (CPET). Since 250° F (121° C) is the maximum range of a water-temperature controller, electric heating elements can be installed to further increase temperature. A standard fluid-cooled mold base also can accommodate heating elements drilled at a different plane level for multiple-

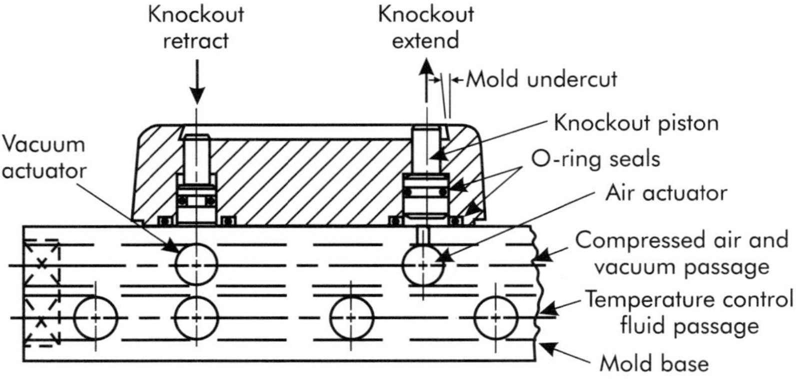


*Figure 8-21. Mold base with two separated passages, one for temperature control and the other to supply air or vacuum to knockout actuators.*

purpose temperature service—as long as both systems are not activated simultaneously.

## Hole Spacing and Size

Fluids or compressed air, and heating elements, require long holes drilled through the mold base length. These through-holes are most often gun-drilled quickly and accurately into the entry face using a specialized gun-drilling tool. Alternately, the operator drills two holes, one from each parallel face, which meet in the center of the plate. Holes should be placed with a minimum wall thickness of 0.18 in. (4.6 mm) apart to avoid breaking through any crossing perpendicular fastener or cavity holes. Fluid passage holes should be a minimum of 1.375 × drill-diameter distance from any face of the mold base and 2 × drill-diameter to adjacent parallel holes. Electrical heating elements inserted into through-holes are fastened on one end only by a set screw because they increase in length considerably when hot and should not be constrained on both ends to avoid damage.

The diameters of the fluid and compressed air passages are often specified using standard pipe tap-drill sizes so that entrances and exits can be sealed with standard pipe plugs, as detailed in Table 8-3.

When a mold is unable to achieve an economical cycle, it is important to investigate if the cavity temperature is far above the fluid-outlet temperature and heat is not properly transferring to the mold base. The problem may reside in the heat-transfer design or fabrication of the cavity due to its height or depth.

Table 8-3. Pipe sizes commonly used for mold bases

| Pipe Size, in. (mm) | Tap Drill Diameter, in. (mm) |
|---|---|
| 1/8 (3.18) | 0.339 (8.61 mm) |
| 1/4 (6.35) | 0.437 (11.10 mm) |
| 3/8 (9.53) | 0.562 (14.28 mm) |
| 1/2 (12.70) | 0.719 (18.26 mm) |

## Fluid Distribution

Fluid distribution in the mold base can be either in series or parallel circuits similar to electrical circuits. Advantages and disadvantages of the two systems are discussed in Chapter 6. Many designers choose the series path for large molds and provide multiple circuits that emanate from a manifold, supplying sufficient flow to maintain a turbulent stream through mold-base passages (see Figure 8-22). A parallel-flow circuit does not require adjusting flow valves at the manifold to achieve an overall temperature gradient across the face of the mold base. However, parallel flow may not develop enough velocity to become turbulent throughout the passages. It may best be suited to moderate thermoforming cycles and where sufficient fluid flow cannot be supplied to a series circuit.

The series fluid circuit requires internal plugs to direct flow between parallel-drilled holes in a straight-line path (see Figure 8-23). Commercial pressure plugs are available from mold suppliers.

Parallel fluid circuits require that the input flow be equally divided among all system water lines (see Figure 8-24). The method used to balance flow can be a restrictor assembly in each water-

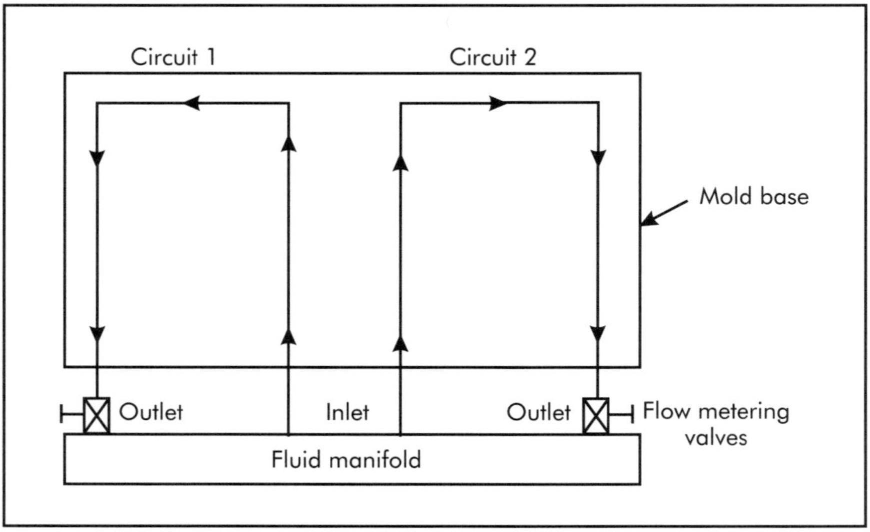


*Figure 8-22. Multiple series fluid circuits emanating from a manifold.*

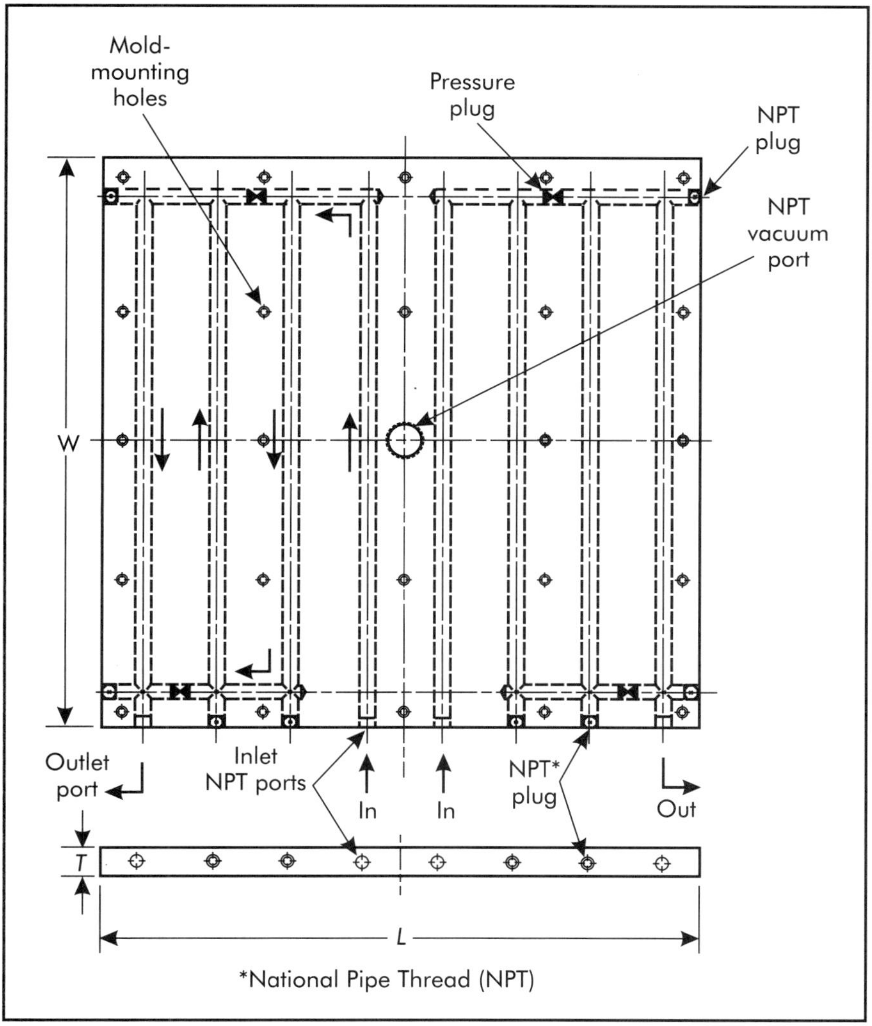


*Figure 8-23. Series fluid circuit using a pressure plug to deflect flow through the mold base. (Courtesy E.D. Segen & Co., LLC)*

line with a clearance area that equals the water-inlet area when multiplied by the number of holes (see Figure 8-25). Equation 8-4 is used to calculate the diameter needed to restrict flow.

$$A_1 = N(A_1 - A_2) \tag{8-4}$$

where:

$$A_1 = \text{water inlet area} = \frac{\pi D_1^2}{4}$$

$$N = \text{number of parallel holes}$$

$$A_2 = \text{restrictor area} = \frac{\pi D_2^2}{4}$$

so,

$$\frac{\pi D_1^2}{4} = \frac{\pi}{4}\left(D_1^2 - D_2^2\right)N$$

$$D_2 = \sqrt{\frac{ND_1^2 - D_1^2}{N}} = \text{diameter of restrictor to equalize flow}$$

in all parallel holes ($N$)

For example:

$D_1 = 0.437$ in. (11.1 mm)
$N = 4$

$$D_2 = \sqrt{\frac{4(0.437)^2 - 0.437^2}{4}} = \sqrt{\frac{3(0.437)^2}{4}} = 0.378 \text{ in. (9.6 mm)}$$

Female cavities may be flood-cooled by machining an annulus on the cavity and sealing the ends with O-rings to avoid fluid leakage from open mold-base faces (see Figure 8-12). A series of these cavities can be interconnected by a waterline designed for turbulent flow (see Figure 8-26). The required area of the inlet waterline is:

$$W_a = (P_w \times L) \times 2 \text{ (paths)} \tag{8-5}$$

where:

$W_a$ = required area of the inlet waterline to induce efficient heat transfer by means of turbulent flow, in.$^2$ (mm$^2$)
$P_w$ = fluid passage width, in. (mm)
$L$ = fluid passage length, in. (mm)
$P_w \times L$ = half of projected area of fluid passage, in.$^2$ (mm$^2$)

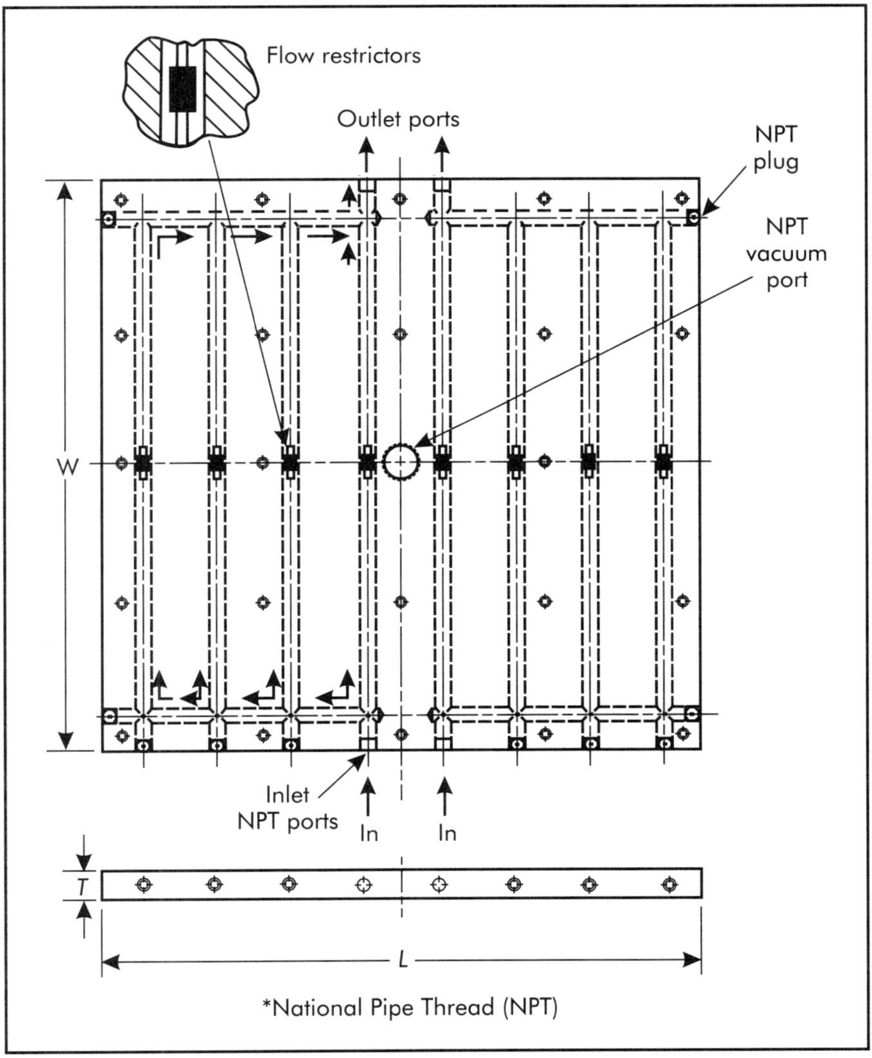


*Figure 8-24. Parallel circuit mold base flow. (Courtesy E.D. Segen & Co., LLC)*

## Molds for Contact Heat

All contact-heat thermoforming molds are female because conduction, rather than a combination of radiation and convection (ovens), is used to transfer heat to the plastic sheet, as discussed

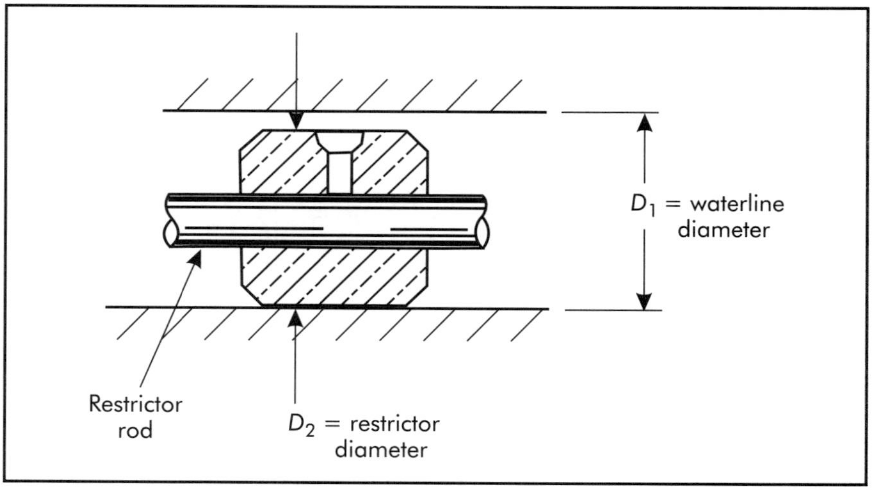


*Figure 8-25. Restrictor rod geometry in a parallel flow mold passage.*

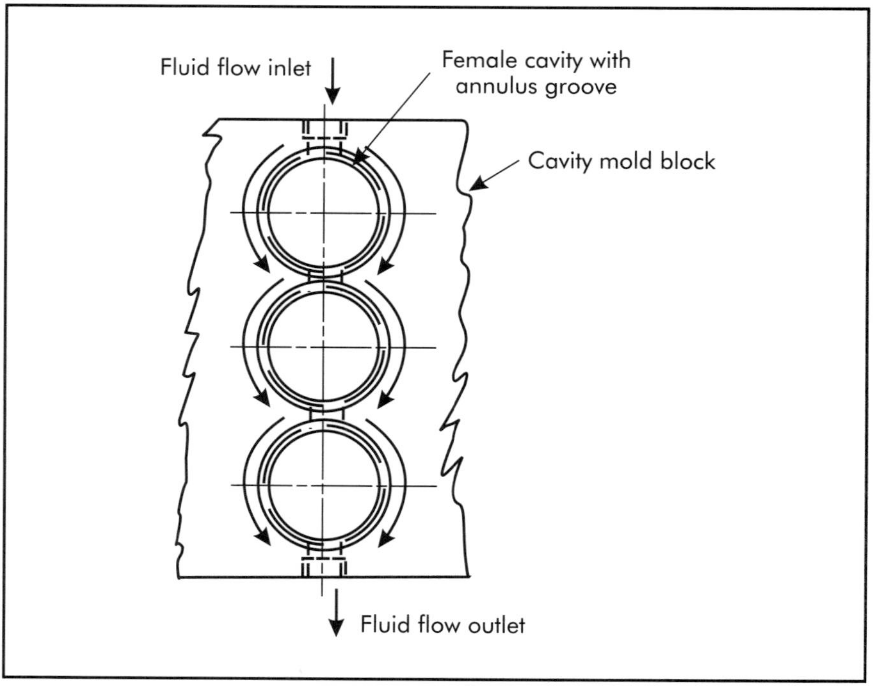


*Figure 8-26. Flood cooling of female molds through an annulus groove around the cavity.*

in Chapter 4. These cavities have a knife-like die encircling their perimeter that provides a seal for pressure forming and a means of trimming the part when the forming cycle is completed (see Figure 8-27). The knife-like die has its sharp edge ground so that a 0.002–0.004 in. (0.05–0.10 mm) flat is produced to act as a seal for the pressure-forming air without cutting entirely through the web. A subsequent press action completes the trim.

A knife edge wall is always incorporated into the product when a flush cut-edge finish is desired (see Figure 8-28). The term *nearly flush cut* is a better description because the knife edges extend beyond the mold and the resulting flange projection can be felt with a fingernail.

## Part Removal from the Mold

Contact-heat undercut parts can be difficult to remove from the mold with air ejection alone because formed parts are cut free from the web at completion of the cycle and held for transport only by partially cut out carry tabs (nicks). Various methods are used to partially cut into the web to allow transport of the formed parts out of the mold. These nicks are located on the sides parallel to indexing, not in the front or back (see Figure 1-14). The best

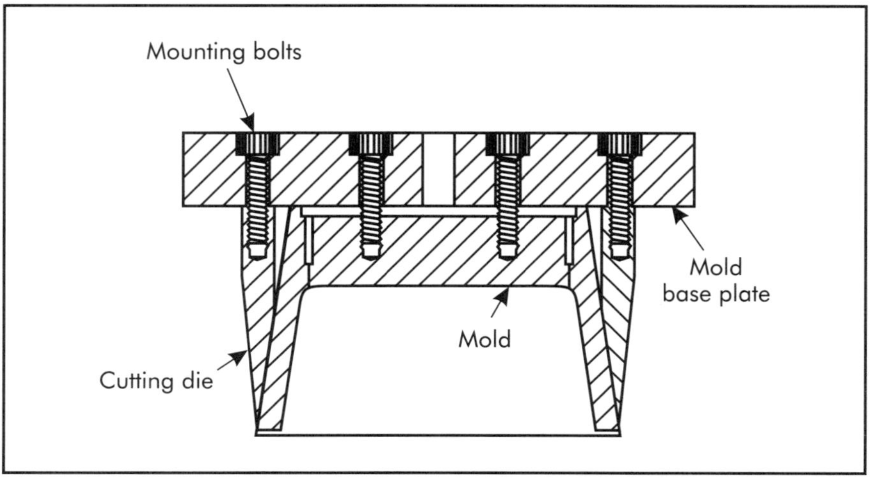


*Figure 8-27. Cut-in-place contact-heat thermoforming tool (Mitchell 1998).*

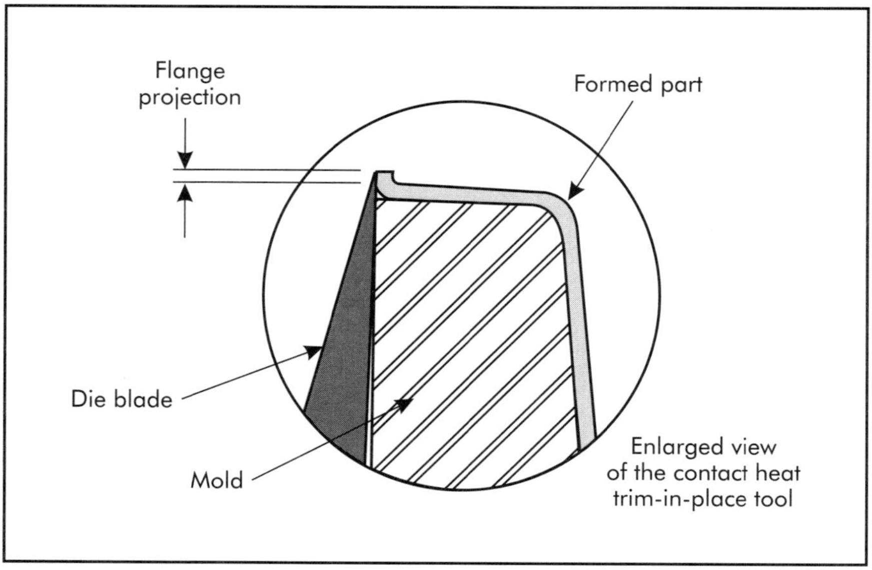


*Figure 8-28. Nearly flush cut-edge made by a cut-in-place contact heat tool. (Courtesy G.N. Plastics Co., Ltd.)*

time to locate and install the nicks is when the tool is set up and being tested in the thermoformer. One method of installing nicks is to dull the knife in a small circular arc at several locations, slowly deepening the nick after making test shots. If the nicks are cut too deep, the formed parts may remain in the cavity when the web is indexed away from the mold.

Spring-loaded knockout pins, located close to the undercut section, can help eject undercut parts. Often the mold flange area is very narrow and the size of the pins must be quite small, 0.125–0.187 in. (3.18–4.75 mm) diameter, so that several pins may have to be grouped together to facilitate smooth ejection. A more elaborate knockout system using a blade supplies more consistent power for ejection of an undercut (see Figure 8-29).

## Layout Requirements

The layout parameters of contact-heat molds are somewhat different from the standard, oven-preheat-style mold.

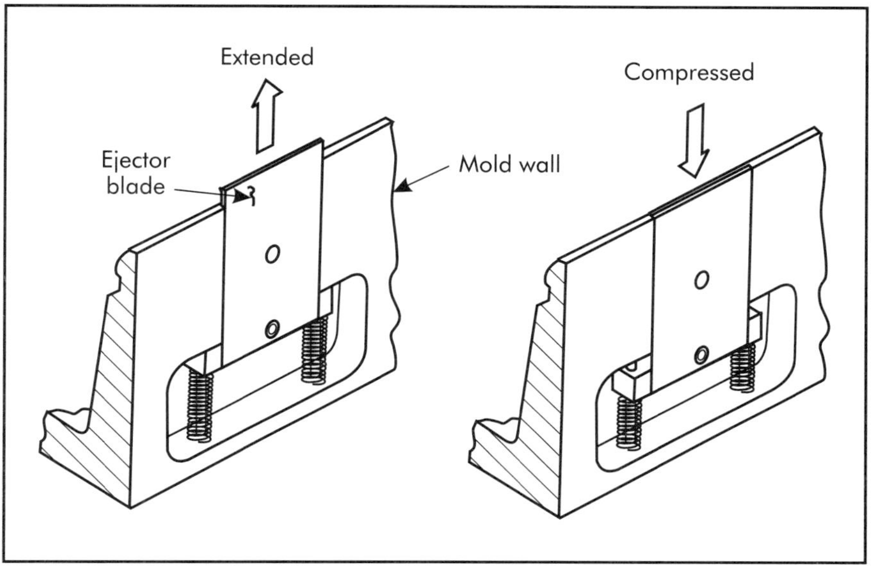


*Figure 8-29. Contact-heat mold ejector blade. (Courtesy G.N. Plastics Co., Ltd.)*

- The contact-heat thermoformer does not use a chain indexing system and only requires a grip of 0.63 in. (16.0 mm) of waste on each web edge to transport shot out of the mold.
- Layout of cavity-to-cavity centers needs only to consider the thickness of the abutting die walls as the minimum distance. A single, common knife can be used to seal off two cavities for suitable straight-wall dies, which will reduce shot width or length. These common knife-cut parts require nicks to support the shot while being transported away from the mold.
- Contact-heat pressure forming cannot use plugs to help distribute hot plastic uniformly throughout the mold. When using female molds, machine manufacturers recommend a ratio of 0.65 between the depth of the mold and its minimum open-top width. This ratio, found by Equation 8-6, avoids causing thin-wall part bottoms on the formed part.

$$D = 0.65W_m \tag{8-6}$$

where:

$D$ = mold depth, in. (mm)
$W_m$ = minimum mold width, in. (mm)

- No allowance need be made for a heavy-wall pressure box, since a welded, forged die serves each cavity as its own personal pressure box. It would be prudent to calculate force on the die sidewalls for long or very deep dies.
- The contact-heated blow plate requires a dedicated baffle of aluminum or copper plate to direct pressure-forming air only to each cavity (see Figure 8-30). The cut-plate heated surface contains small, drilled holes located on 0.50 in. (12.7 mm) centers. The baffle plate must direct airflow only within this mold area and efficiently transmit heat from the hot blow-plate source to the cut-plate face. Holes in minimum numbers are drilled through the baffle plate to avoid reducing the surface conducting heat to the blow plate.
- Mold cavities are fastened to the mold plate or base in the usual manner for best heat transfer, but the die may be fastened by several methods (see Figure 8-31). Die makers may tap the narrow wall blade with small-diameter threaded holes

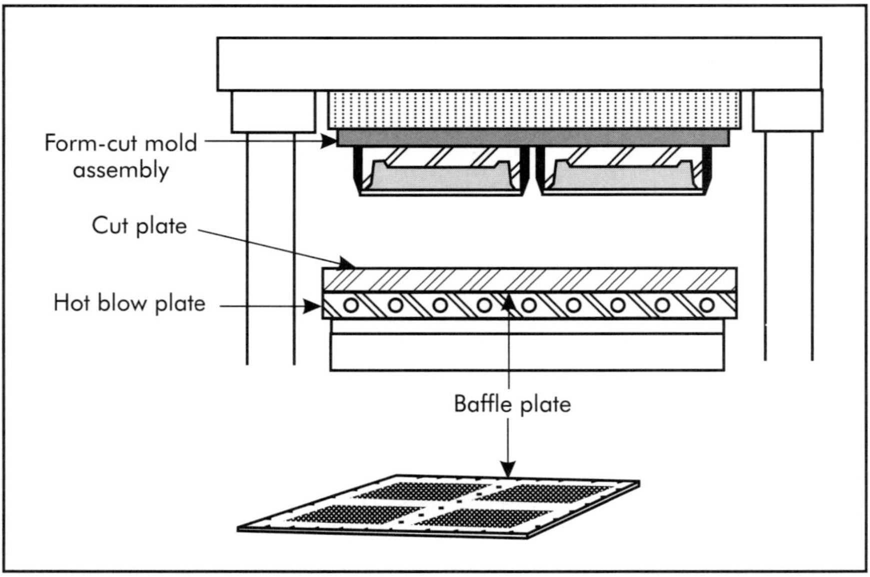


*Figure 8-30. Baffle plate installation. (Courtesy G.N. Plastics Co., Ltd.)*

on the mold plate (#1 method). If space is sufficient on the outer perimeter of any two parallel sides of the die, a short outer strut or tab can be welded on the die during its fabrication to bolt the die in place (#2 method). When the outer space is limited, a narrow inner strut or tab can be welded to secure the die in place (#3 method). The internal welded member requires that the mold back face provide clearance for the strut. This reduces heat transfer to the cooling plate by reducing the mold area in contact with plate. The welded struts should be ground flush with the back of the die during manufacture so they are in contact with the mold base.

## Conclusion

Cavity layout and mold-base design are interlinked by many factors. Paramount is maintaining uniform cavity temperature and efficient heat transfer for fast cycling. The mold purchaser requires the maximum number of cavities per shot to reduce sheet waste and minimize thermoforming machine running time. Mold

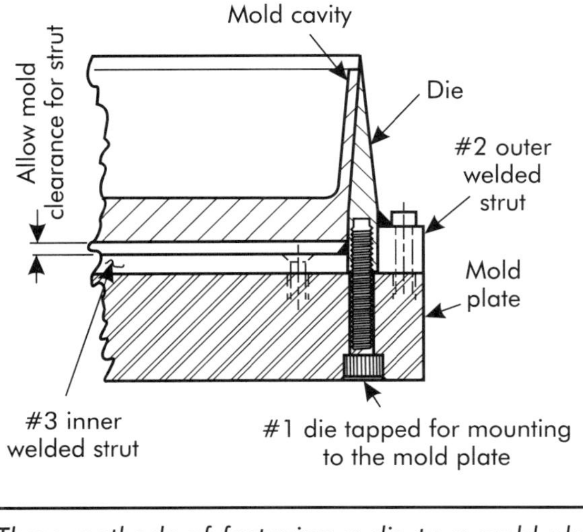


*Figure 8-31. Three methods of fastening a die to a mold plate. (Courtesy G.N. Plastics Co., Ltd.)*

designers have to decide on the best of many conflicting options to keep tooling cost reasonable, yet be certain that the completed mold functions without unpleasant surprises. Mold builders work best when the design is clean and straightforward and their input is considered and valued.

Figure 8-32 shows specifications that can prove useful to both the customer and mold designer.

## Reference

Mitchell, Philip, ed. 1998. *Tool and Manufacturing Engineers Handbook*, Fourth Edition, *Volume 8: Plastic Part Manufacturing*. Dearborn, MI: Society of Manufacturing Engineers.

### Checklist of Mold Details

1. Plastic resin specified ______________ Plastic shrinkage______________
2. Critical wall thickness ___________________ Location ______________
3. Critical fit area as per prototype testing data _______________________
4. Engraving _______________ Texturing ___________________________
5. Thermoforming machine model and manufacturer _________________
6. Thermoform equipment specifications:
   A) Pressure form _______________ Vacuum form _________________
   B) Maximum mold size: width _____ length_____ height _______
   C) Roll-fed thermoformer with oven preheat web ___________________
   D) Contact-heat thermoforming __________________________________
   E) European mold and trim-in-place system ______________________
7. Trim press model and manufacturer_______________________________
   Type of die
   A) Steel-rule die _______
   B) Forged knife-like die _______
   C) Punch and die _______
   D) Contact-heat tool die ___________
   E) European trim-in-place die _______
8. Number of cavities per shot specified by user_______
9. Cavity material:
   A) Cast plastic __________
   B) Cast aluminum______________
   C) Machined aluminum _______
10. Mold finish:
    A) Sand or shot blast: rough ______ medium _____ fine ______
    B) Polish: fine _______ medium _______ coarse _______
    C) Teflon™ coat _______
    D) Other ____________________________________
11. Type of mold cavity:
    A) male ____ female ____ foam mold ____ contact heat ____
       other ____________________
    B) Family style to allow interchangeable cavities _______
    C) Cavity mounting style:
       a) Cavity mounted directly onto a dedicated temperature-controlled mold base_______
       b) Cavity mounted onto a mold plate that will be mounted to an existing temperature-controlled mold base _______
       c) Cavity to be internally drilled for liquid temperature passages and mounted to a dedicated mold base _______

*Figure 8-32. Checklist of mold details.*

12. Temperature-controlled mold base specified:
    A) New, dedicated to one style of cavity ________
    B) New mold base that can be used for many mold plates of same size ________
    C) New, adjustable-length mold base for many mold plates of equal width and variable length ________
    D) Customer will provide an existing mold base for this mold ________
13. Pressure box:
    A) New, single-purpose dedicated pressure box _____
    B) New pressure box to be used with a variety of other molds that will fit in the box _____
    C) New adjustable-length pressure box to be used with a variety of other molds of equal width and variable length _____
    D) Customer will provide an existing pressure box _____
14. Assists for male molds:
    A) Ring assist ____
    B) Simple assist ____
15. Plugs for female molds:
    A) Electrically heated and temperature controlled _____
    B) Hot oil and temperature controlled _____
    C) Insulated material to be used for unheated plug:
        a) Wood ______
        b) Wood covered with felt______
        c) Epoxy______
        d) Syntactic foam ______
16. Air cylinder sheet clamps ______ Gas-activated spring clamp ______
    A) Mounted on mold base ______
    B) Mounted on pressure box ______
    C) Front and rear only ______
    D) Four-sided clamp ______
17. Quick-change tooling system manufacturer:________________
18. Mold undercuts __________
19. Ejecting completed shots:
    A) air blow-off ____
    B) air-cylinder activated ____
    C) thermoforming press knockout activated ____
    D) knockout plates ____
    E) knockout pins ____
    F) articulating mold section ____
    G) other ________________
20. Special features:________________________________

*Figure 8-32. (continued)*

# 9

# Mold Cost Estimating

This chapter will discuss mold cost estimating and variables that affect profit. Forms are included to make the job of mold estimating easier.

## Factors that Make Up Cost

A thermoforming mold supplier often is called upon to quote a fixed price to a potential customer. If the price is low, the mold maker will incur a financial loss, and if the quote is too high and noncompetitive, the order will be lost. The seller must accurately estimate both direct and indirect costs, plus include a reasonable profit to arrive at a fair selling price.

### Indirect Costs

Indirect costs can be grouped into two categories: variable and fixed. Variable indirect costs may include the costs of:

- sales,
- advertising,
- trade shows, and
- travel and entertainment.

Fixed indirect costs may include the costs of:

- business office and management personnel,
- plant real estate,

- plant utilities,
- company vehicles,
- taxes and social security,
- insurance,
- pension and vacation periods, and
- computers and shop equipment.

Indirect costs can be compiled from the previous year's statements and updated by an accountant. The burden can be distributed on an apportioned hourly basis among appropriate personnel of the company as overhead cost-per-hour. The author does not attempt to list all indirect costs, but does provide a list of direct-cost factors to assist the mold-estimating process.

## Shop Backlog

The percentage of profit per project fluctuates, depending on the size of the potential order when compared to the existing order backlog, the capacity available for design and fabrication, and the expected tool complexity. Companies that have a small backlog of work may accept tough projects at low profit margins because their indirect costs continue to mount day after day. Alternately, a fat project backlog tends to increase the price of difficult tooling, which may absorb too much time and energy from skilled personnel.

A small minority of customers are impossible to satisfy either because they require a political committee to approve every mold detail, or because an individual buyer by nature is inherently negative. It is wise to ferret out these troublemaking end users and avoid them because they never will allow a mold supplier to make a fair profit.

On-time delivery of molds may demand unexpected excessive shop overtime work at premium wages to meet commitments made to a customer. Unforeseen design or fabricating costs have to be borne by the firm producing the mold. From time to time, finished molds or cavities must be reworked, and those changes cannot be passed on to the customer. The costs must be absorbed by the toolmaker. A contingency factor based on past experience can cover these losses and maintain a fair overall profit.

## Direct Mold Costs

Direct mold costs consist of the dollar value of purchased materials, perishable tools, and the direct cost of all labor. Commonly purchased materials may be ordered in larger quantities than necessary. The extra material costs less as part of a larger order and can be kept for future use. The value of material used directly for the mold can be based on a nondiscounted price and should include cutoff scrap, which occurs whenever standard stock sizes are ordered.

The direct cost of labor can be kept simple by using the actual labor hourly rate and benefits, plus a proportional share of indirect costs. A premium may be charged for items machined on very expensive machine tools. Estimated labor consists of the number of estimated hours times the appropriate direct hourly cost for each task.

The estimated mold cost can be made more accurate if the number of cost factors is broken down into small fragments and each is assigned a cost. If a serious estimating error is made on one fragment, the other items often compensate to balance out pricing. In the final analysis, the mold estimate should be reviewed and studied to see if the final price makes practical sense when compared to similar completed molds.

## Variables that Affect Profit

When all direct and indirect costs have been tallied in the estimate, a standard profit percentage is added to the total costs. The standard profit percentage does not exist as a textbook figure, but should be a monetary goal that is essential to maintaining and growing a business. Modification of the standard profit may be a necessary strategic business decision, but it cannot be intelligently made until the true total cost has been tallied and reviewed for any flaws.

## Complexity

The complexity of the project judges whether the mold will move smoothly through the shop and be delivered on time. If the project is relatively straightforward and simple, competition may cause the standard profit percentage to be adjusted downward to be acceptable to a customer.

A pioneering and complex mold can tie up an engineering department and mold shop for extended periods of time, disrupting other profitable work and wrecking delivery schedules. This kind of project difficulty requires a company with sufficient resources so that, if the undertaking bogs down, the rest of the organization can continue to function normally. The financial risk involved deserves a rich premium since complex molds are speculative in nature and are undesirable business for most small and mid-sized companies.

The standard profit percentage can be reduced by 3–7% for simple molds and increased by up to 20% for complex molds.

## Profit Contingency

A profit contingency factor is often necessary since unknown losses can occur even on the most mundane of molds. Many times this extra funding is not required for most molds. However, suddenly all the accumulated contingency money may be poured into the next mold. In the experience of some business veterans, at the end of the fiscal year all the contingency funds manage to be spent and none show up in the bottom profit line.

As a percentage of the total quote, it may be prudent to add a minimum of 3–7% to the total mold cost to cover profit contingency.

## Back Orders

When the mold shop has a heavy backlog and is unable to deliver new orders on time, honesty is the best policy. Long-time customers may still insist on a quote and it may be unwise to put them off. If the shop has a continuing relationship with reliable subcontractors, portions of the mold may be sent out for fabrication and internal shop overtime increased. This may create sufficient additional capacity to satisfy the buyer if he is willing to pay a realistic premium above the additional cost for these efforts. However, nothing will destroy a business relationship faster than paying a premium for a late delivery. Thus, careful weighing of the pros and cons in accepting the order is necessary.

## Customer Assistance

A neophyte customer starting up a new thermoforming operation needs a great deal of handholding, which goes far beyond selling him molds. This type of purchaser often requires information about common business practices and thermoforming problems.

## Most Often Asked Questions

### Labor

*Question:* Where can I hire experienced supervisors and operators?

*Answer:* I can not help you locate currently employed skilled employees. The labor market is very slim so that all new employees may have to be trained internally.

### Thermoforming Parts Formed Improperly

*Question:* Are improperly formed parts caused by the mold, machine, plastic-sheet problems, or lack of thermoforming skills?

*Answer:* Our mold will form the part properly but we will assist and advise you to the best of our ability to get production started. This unpaid consulting work can be the start of a business relationship lasting many profitable years if the mold supplier is knowledgeable and willing to invest the necessary time to help a novice who is capable of acting on the basis of counsel.

## Uneconomical Initial Quantity of Plastic Formed Parts

*Question:* I am new to the industry and my customer's initial quantity orders are small. How can I buy a mold made within my budget?

*Answer:* I can quote on a smaller number of less sophisticated cavities than would be normally recommended to lessen tooling costs. Cost for your labor, machine time, and plastic materials may tend to exceed estimates but will allow proceeding with your initial order.

Machinery and plastic suppliers also have a responsibility to help get a new thermoforming operation up and going. If they are not technically knowledgeable, they may shift responsibility for proper functioning away from their equipment. Technical assistance from the mold supplier to the novice at this critical time can help build self-confidence in the novice and faith in the mold maker.

## Mold Estimating Forms

The mold estimating form shown in Figure 9-1 may not contain all the items that pertain to every mold. Users may wish to add suitable items to fit a particular mold design. A customer can be very reluctant to approve additional payments for items missed during the estimating process, so it is important to check the original specifications for all details. Breaking an estimate into tiny bits and putting a cost to every detail increases confidence in the overall final quote.

The estimating form provides direct costs only. A fair price must include a profit percentage added to these costs.

**ESTIMATE OF DESIGN AND ENGINEERING DIRECT COST** QUOTE NO. ________
**QUOTATION SHEET**

Date: ________ Est. by: __________ Page ________ of ________

| Part Description | Hours | × Hourly Rate = | Cost | Remarks |
|---|---|---|---|---|
| **DESIGN, TESTING, AND PART DETAIL DRAWING** | | | | |
| Preliminary product design, modification, and approval | | | | |
| Prototype mold design (male, female, plugs) | | | | |
| Testing, modifying prototype mold | | | | |
| **MOLD ENGINEERING DECISION MAKING** | | | | |
| Temperature control system, mold base cavities, and stripper plates | | | | |
| Number of cavities and preliminary spacing on mold base | | | | |
| Fixed- or adjustable-length pressure box and mold base | | | | |
| Reviewing customer's special needs | | | | |
| Knockout and sheet clamp system | | | | |
| Cavity fabrication—casting or machining | | | | |
| Method of thermoforming uniform plastic walls (plugs or assists) | | | | |
| Reviewing trim system to confirm compatibility to mold | | | | |

*Figure 9-1. Mold estimating form.*

Page ________ of ________

| Part Description | Hours | × Hourly Rate = | Cost | Remarks |
|---|---|---|---|---|
| **MOLD ENGINEERING DECISION MAKING** | | | | |
| Special cavity finishes or engraving | | | | |
| Specifying any special purchased parts required | | | | |
| Review all thermoforming machinery manufacturers' data to confirm that mold is compatible to the machine | | | | |
| Plug heating methods and materials specified | | | | |
| **MOLD LAYOUT, DESIGN, DETAIL** | | | | |
| Cavity layout on mold base | | | | |
| Cavity drawing | | | | |
| Mold base layout (fixed or adjustable) | | | | |
| Mold base buildup to sheet line | | | | |
| Detail drawings of individual mold parts | | | | |
| Pressure box layout | | | | |
| Pressure box layout to sheet line (fixed or adjustable length) | | | | |
| Plug or assist layout | | | | |

*Figure 9-1. continued.*

Page ________ of ________

| Part Description | Hours | × Hourly Rate = | Cost | Remarks |
|---|---|---|---|---|
| **MOLD LAYOUT, DESIGN, DETAIL** | | | | |
| Plug heating design, if required | | | | |
| Pressure box detail | | | | |
| Purchased parts installation drawings | | | | |
| Electrical layout | | | | |
| Plumbing layout | | | | |
| Positioning and locating devices for quick mold installation | | | | |
| Check and review design and details | | | | |
| **CAVITY MACHINING AND FINISHING*** | | | | |
| Interior cavity machining | | | | |
| Exterior cavity machining | | | | |
| Cavity interior finish | | | | |
| Vacuum hole drilling | | | | |
| Vacuum distribution grooves | | | | |
| * Cost per cavity × number of cavities + spares | | | | |

*Figure 9-1. continued.*

Page ________ of ________

| Part Description | Hours | × Hourly Rate = | Cost | Remarks |
|---|---|---|---|---|
| **CAVITY MACHINING AND FINISHING** | | | | |
| Internal water passages for cavity cooling (if required) | | | | |
| Mounting holes to mold base | | | | |
| Special additional work (for example, engraving) | | | | |
| **MOLD BASE MACHINING** | | | | |
| General machining interior and exterior of plates | | | | |
| Female mold base cavity machining (if required) | | | | |
| Drilling and tapping of fastener holes | | | | |
| Vent hole drilling and machined vent passages | | | | |
| Cavity mounting holes | | | | |
| Drilling and tapping of water passages and electrical heating elements | | | | |
| O-ring water inlets for cavity water cooling (if required) | | | | |
| Vacuum vent collection plate (if required) | | | | |
| Press platen to mold base mounting plate | | | | |
| Knockout mounting holes | | | | |

*Figure 9-1. continued.*

Page ________ of ________

| Part Description | Hours | × Hourly Rate = | Cost | Remarks |
|---|---|---|---|---|
| **MOLD PLATE AND CAVITY ASSEMBLY MACHINING (if specified)** | | | | |
| Mold plate machined to size | | | | |
| Mold plate cavity mounting holes | | | | |
| Mold plate machine vent passages | | | | |
| Perimeter wall for female mold (if required) | | | | |
| Mold plate mounting holes and threaded insert | | | | |
| O-ring water inlets for cavity water cooling (if required) | | | | |
| **PLUG AND ASSISTS** | | | | |
| Plug machining and mounting holes | | | | |
| Assist assembly | | | | |
| *Note: Plug heating, if needed, can be complex and should be estimated as a separate project.* | | | | |
| **SHEET CLAMPS** | | | | |
| Four-sided or front and back clamps | | | | |
| Cylinder brackets and associated items | | | | |
| Air cylinders or gas springs | | | | |

*Figure 9-1. continued.*

Page ________ of ________

| Part Description | Hours | × Hourly Rate = | Cost | Remarks |
|---|---|---|---|---|
| **SPECIAL MOLD BASE FEATURES** | | | | |
| Off-line trim index pusher tabs | | | | |
| Knockout plates and pins | | | | |
| Mold articulating systems | | | | |
| Quick-mounting change locks (if required) | | | | |
| Special water distribution manifold | | | | |
| Special measures | | | | |
| **PRESSURE BOX MACHINING** | | | | |
| Fixed-size pressure box back plate | | | | |
| Adjustable-length pressure box back plate | | | | |
| Sidewalls | | | | |
| Air baffle | | | | |
| Plug or assist mounting plate | | | | |

*Figure 9-1. continued.*

Page ________ of ________

| Part Description | Hours | × Hourly Rate = | Cost | Remarks |
|---|---|---|---|---|
| **PRESSURE BOX MACHINING** | | | | |
| Spacer bars (number required) | | | | |
| Platen mounting plate | | | | |
| Mounting quick-change locks (if required) | | | | |
| **ASSEMBLY OF COMPLETE TOOLING** | | | | |
| Mechanical assembly of machined parts | | | | |
| Plumbing of fluid passages and air and vacuum connections | | | | |
| Electrical assembly of heating elements, sensors, and switches | | | | |
| Testing knockouts, articulating components, and clamp frame | | | | |
| Plastic thermoforming test of mold to check process and finish parts | | | | |
| **Part Description** | **No. of Units** | **× Cost/Unit =** | **Total Cost** | **Remarks** |
| **MOLD RAW MATERIAL COST** | | | | |
| Cavities: | | | | |
| *A) Cast aluminum cavities | X | X | X | X |
| Casting pattern (1) | | | | |

*Figure 9-1. continued.*

Page ________ of ________

| Part Description | No. of Units | × Cost/Unit | = Total Cost | Remarks |
|---|---|---|---|---|
| **MOLD RAW MATERIAL COST** | | | | |
| As-cast cavity | | | | |
| Rough casting cleanup | | | | |
| **OR** | | | | |
| *B) Machined aluminum cavity/raw stock | | | | |
| C) Plug raw material | | | | |
| D) Assist raw material | | | | |
| * *Choose one of items A or B. Allow a scrap minimum of one extra cavity plus an additional spare for every 10-12 cavities required.* | | | | |
| **MOLD BASE ALUMINUM PLATES** | | | | |
| Mold base estimate weight ___________ | | | | |
| Mold plate estimate weight ___________ | | | | |
| Mold spacer, and related items _______ | | | | |
| **PRESSURE BOX MATERIALS** | | | | |
| Back plate aluminum | | | | |
| Sidewalls | | | | |

*Figure 9-1. continued.*

Page ______ of ______

| Part Description | No. of Units | × Cost/Unit | = Total Cost | Remarks |
|---|---|---|---|---|
| **PRESSURE BOX MATERIALS** | | | | |
| Pressure box buildup | | | | |
| **MISCELLANEOUS ITEMS** | | | | |
| Fasteners and pins | | | | |
| Knockout materials | | | | |
| Surface finish (plating, black oxide, sandblast) | | | | |
| Valves | | | | |
| Plumbing supplies | | | | |
| Electrical supplies and heating elements | | | | |
| Seals | | | | |
| Indexing buttons | | | | |
| Quick-change locks | | | | |
| Specially shaped machine-shop cutters | | | | |
| Perishable cutters | | | | |

*Figure 9-1. continued.*

# 10

# Knife-like Trim Dies

Knife-like trim dies are readily available at relatively low cost and with rapid delivery from many different sources. The best die suppliers are those who have experience dealing with specific problems common in the thermoforming industry. Other die makers specialize in serving the graphic arts and box-making industries and, if they are to be engaged to build thermoforming dies, may require considerable advice and guidance from the die user.

There are three basic types of knife-like dies: steel rule, forged (high die), and machined-tool-steel knife edges (see Figure 10-1). All of these trim blades are sharpened to a keen edge, are of precision height and parallelism, and conform to the shape to be trimmed. They vary in maximum practical knife height and trim-life longevity based on tool-steel composition and the maximum hardness of the cutting edges. Fabrication techniques unique to each type of die will dictate price, delivery, and quality.

## Steel-rule Dies

Steel-rule dies are fabricated from a ribbon of prehardened, sharpened steel rule. The thickness is measured and specified in *points* (a printing term). Thermoforming dies are generally manufactured of 3 point (0.042 in. [1.07 mm]) rule. Standard steel rule must be ductile enough to bend into tight radii, so it is moderately hardened to only 35 Rockwell-C ($R_C$) or a bit higher. Other types of rule are available with a relatively hard edge (45 $R_C$ or higher) and soft ductile body. This type of rule may be used when

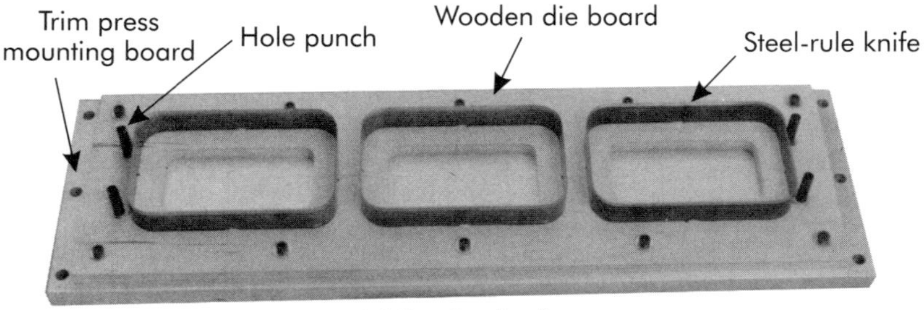


(a) Steel-rule die

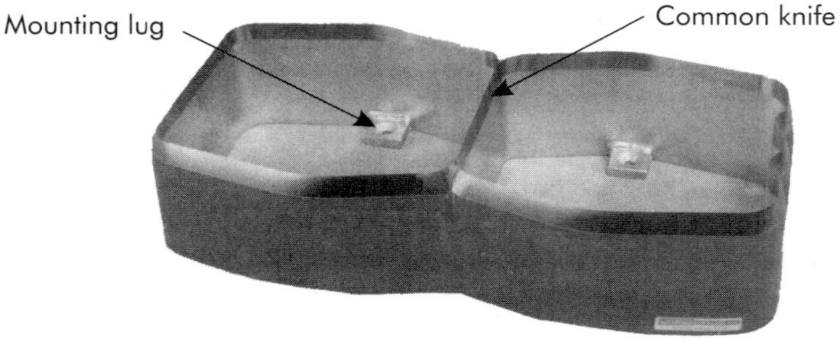


(b) Forged die cavity

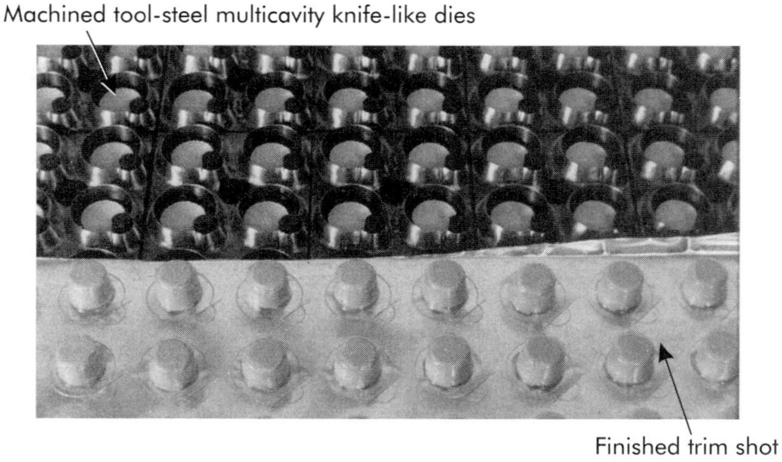


(c) Machined tool-steel knife-like die

*Figure 10-1. Three types of knife-like dies. (Courtesy E.D. Segen & Co., LLC and Ontario Die Ltd.)*

forming a small radii, which otherwise may fracture the metal at the bend area. Most commercial rule is available in heights of 0.937–3.000 in. (23.80–76.20 mm), with many standard sizes in between these limits.

## Die Blade Types and Uses

The sharpened edge can be ordered with a variety of ground angle bevels that gradually reduce rule-body thickness to the required cutting geometry (see Figure 10-2). A single (or side) bevel is ground on one face and a center bevel is ground on both faces. The bevel can be blunt (not good for cutting plastics), have a long acute angle (best for plastics), or be a double-double bevel, which thins the cutting-edge cross section considerably and reduces required cutting force. Perforating rule, which has its knife-edge cut away in a regular pattern, can be used to weaken a package area that is later torn open by the consumer.

A single-bevel knife can be bent so the cutting edge abuts either the interior or exterior of a die cavity. When the cutting edge is adjacent to the interior, it can be used to trim as close as possible to a thermoformed wall and the resulting edge is nearly flush. The plastic trimmed edge can never be flush to the wall because all single-bevel knife edges have a slight chamfer (0.015 in. [0.38 mm]) on the blade, opposite to the main bevel, for sharpening purposes. When two formed cavities are very close together, the die can be designed so the cutting edges of the dies are on the exterior wall of the die with the blade thickness and bevel area facing the inside of the cavity. A center-bevel knife is applicable when two

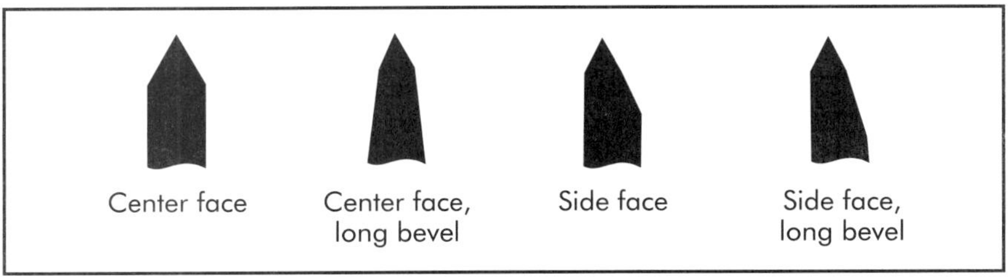


*Figure 10-2. Typical steel-rule cutting edges. (Courtesy J.F. Helmold Corp.)*

cavities are to be split apart with one blade. Perforating rule can be substituted for any of these conditions. However, it is prudent to test a single-cavity die before building a multicavity perforating die. The number of perforations per inch of rule can be varied to achieve the correct weakening of the tear zone.

## Die Board Selection

Steel rule is pressed into laser- or saw-cut slots in a very high-grade laminated wood die board, which acts as the base of the die. Standard die board is fabricated from laminated birch layers, but premium board is made from a hard maple laminate, which is very dense and stable. These boards are available in 0.50–1.00 in. (12.7–25.4 mm) thickness, and their principal purpose is to maintain the rule square to the base. Very accurate dies should have a maximum 0.30 in. (7.6 mm) extension of steel rule above the board. Sometimes several die boards are bolted together to maintain a minimum of blade extension. The die board may be cut out internally to clear the thermoforming part to avoid damage by contact with the board. Thermoformed parts that are deeper than the standard rule can be trimmed on dies or built up on die board or metal pedestals. The pedestals have suitable cutouts to provide clearance for the height of these parts (see Figure 10-3). The back

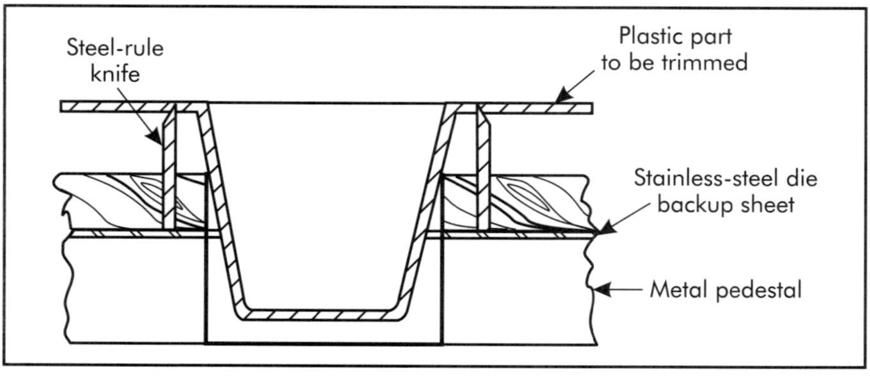


*Figure 10-3. Steel-rule die built on a metal pedestal to increase its depth to clear a thermoformed part.*

face (opposite to the knife edge) of a steel rule should not rest on a bare wood or aluminum pedestal. Rather, it should bear the cutting load directly against a stainless steel sheet of 0.06 in. (1.5 mm) thickness to prevent the back of the rule from being driven down into the softer die buildup.

## Layout Techniques

The layout of the die is drawn directly on the die board when manually jig sawing the slots. For a computer-generated layout, the slot is laser cut directly into the board. The steel-rule slot width is sized to provide a tight friction fit for the rule in the board. If the die becomes damaged or worn, the existing rule can be removed and the die can be re-ruled at a lower cost than fabricating a new die. The layout tolerance when measured between any two cavities is in the order of 0.010–0.015 in. (0.25–0.38 mm) for manual layout and 0.001–0.005 in. (0.03–0.13 mm) for computer layout. Overall accuracy is dependent on the complexity of the die, the rule-bending ability of the die maker, the blade extension from the die board, and other factors that can increase these tolerances.

## Bending and Die Finishing

Steel rule is bent to shape manually or by computer-directed equipment with the intention of producing a die requiring the minimum number of open seams in the rule. The best die is one made from one length of rule containing only one open joint. Any attempt at welding or brazing a joint may change the temper of the rule and reduce the hardness and sharpness of the knife edge. A skillful die maker can spot weld a metal strap lap-joint reinforcement below the knife edge at the seam, creating a nearly continuous knife edge without drawing the temper of the rule (see Figure 10-4). Care must be exercised during spot welding to keep both sharp edges of the joint in a straight plane or the die will not cut cleanly at that juncture. Every unreinforced rule joint tends to open under the force of trimming, exhibiting itself as a

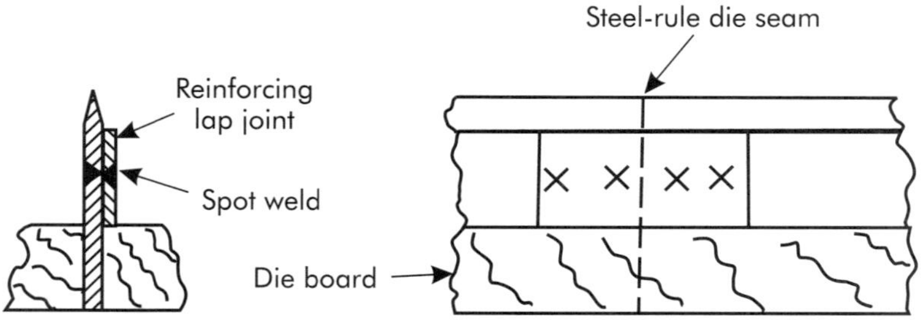


*Figure 10-4. Spot-welded lap joint at a steel-rule seam.*

blemish on the plastic trimmed edge and possibly enlarging the die outline. The best place to lay out a seam in the die is along a straight section of die, never at a radius, which tends to spread open due to internally induced trim forces.

## Stripping of Cut Parts

Trimmed parts often require a stripping action to release them from the die. The simplest and least expensive stripping method is to cement narrow widths of rubber or plastic sponge along both sides of the rule at critical spots to facilitate part and shot skeleton removal (see Figure 10-5). Spring-action stripping plates or bars can be useful for difficult cases of part release and also act as locator pilots for accurate part registration (see Figure 10-6). Care must be exercised to avoid installing an excess of sponge material in the die because it requires a great deal of trim-press energy to compress large areas of sponge, thereby reducing the trimming force available for cutting.

## Buildup for Part Height

When a die requires additional space to clear the depth of a plastic part, a wooden die-board pedestal can be built up under the die board as an inexpensive means of providing clearance. A die board is relatively parallel, but its thickness tolerance is far less precise than a machined or ground metal face. A wooden

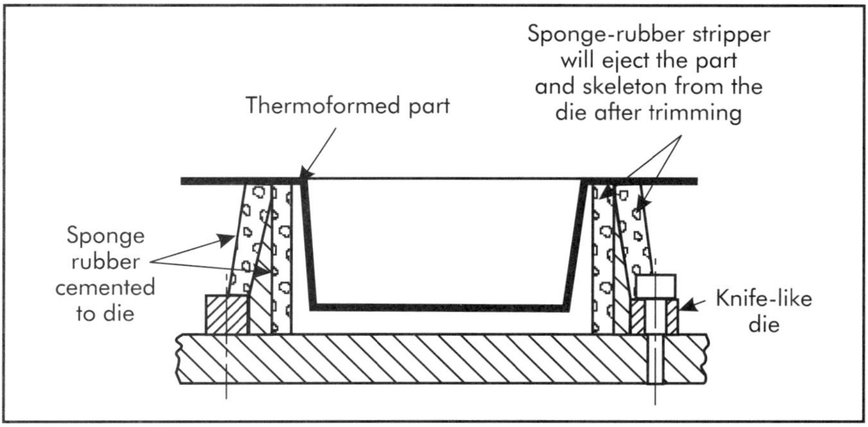


*Figure 10-5. Ejecting trimmed parts and the skeleton from a knife-like die using sponge rubber.*

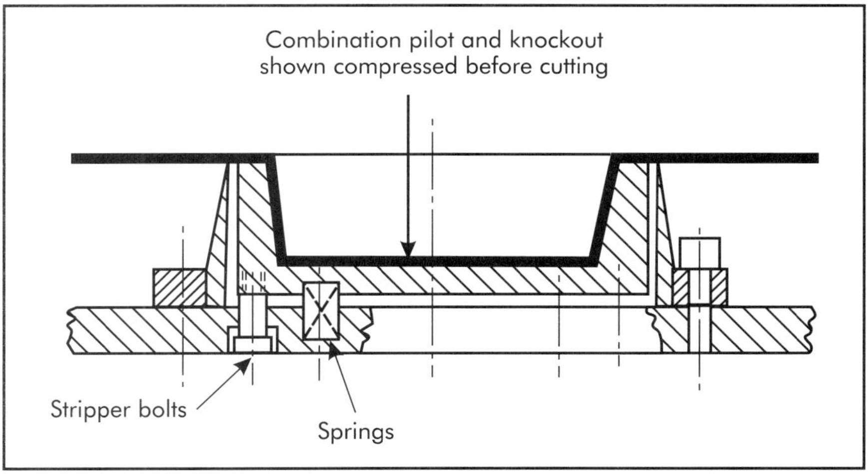


*Figure 10-6. Spring-action combination pilot and knockout locate and eject a trimmed part.*

buildup may call for more extensive manual shimming (make-ready) effort than a machined metal pedestal. The metal pedestal may require cutouts to provide clearance for the thermoformed parts. This can require machining or flame cutting, and final finish-surface grinding. The higher cost of the metal pedestal often can be recouped by faster die setup time at the trim press.

## Tolerances

The tolerance of a completed steel-rule die when measured from cavity to cavity can be at best ±0.015 in. (±0.38 mm) when manufactured by hand layout and manual jig sawing. A computerized layout that directs a laser to burn the board might increase the maximum overall accuracy to ±0.008 in. (±0.20 mm). This will depend on the board's geometry and complexity, and if a minimum amount of rule extends above the die board. The tolerances cited are not average, but are typical of the best of dies. Often, commercial dies may have far looser tolerances.

## Shrinkage

The plastic shrinkage factor can be estimated as one-half of the specified cold shrinkage when an in-line thermoformer indexes to a trim press located next to it on a machine frame (see Chapter 8, Table 8-1). The difference between trimming a cold shot off-line and trimming in-line is that the in-line shots are often still hot and not fully shrunk when indexing into the die-cutting press. Off-line trimming part shrinkage is more predictable because the time lag allows most of the plastic heat to dissipate before die cutting.

## Trimming Solutions

The discussion of tolerance and shrinkage in this chapter focuses on a best effort to trim and register multiple parts accurately and uniformly. If for any reason the result from the initial steel-rule die is not acceptable to a customer, two choices are available:

- Cavities on a standard, multicavity, steel-rule die cannot be moved to compensate for shrinkage, but a new die can be fabricated benefiting from the shrinkage lessons learned from the original die.
- A more expensive die technique using individually mounting floating steel-rule dies to allow each die cavity to shift and pilot itself to the formed cavity can be specified.

An ordinary steel-rule die is usually specified when a minimum quantity of parts is to be trimmed. This type of die is available at the lowest cost with the most rapid delivery, and its overall quality and accuracy are within reasonable expectations. The steel-rule die maker and the customer should know the "rules of the game." The user is obligated to pay for both dies if the first die was built to the estimated shrinkage, but does not register satisfactorily to the shot. The cost to the customer includes production downtime while awaiting delivery of a redesigned die. In-line thermoforming die shrinkage is not an exact science. The ability to float or shift individual dies can be an invaluable asset.

## Forged Dies

Forged (high) dies are bent from unsharpened, bar-tool steel pre-forged into an approximate knife-like edge. This type of die has no open joints since the ends are welded to provide a continuous contour with little chance of changing shape. The unfinished die is heat-treated to 45–56 $R_C$ hardness and ground to the required height, flatness, and parallelism. The die edge is then hand sharpened to the desired trim perimeter and specified bevel. Forged dies can be accurately CNC-machined, providing duplication of each cavity trim-edge shape. Dies deeper than the available standard die stock are fabricated by welding hot-rolled mild steel to the back of the die steel to increase its depth. The two pieces are bent together to form the die. Various methods, as shown in Chapter 8, Figure 8-32, are available to fasten individual dies to a base plate to produce a multi-cavity trim die. The die-mounting plate may be of ground steel or precision aluminum to provide an accurate mounting surface for the dies.

## Die Adjustment

Individual forged dies can be moved from their initial design locations to accommodate the true shrinkage experienced at the trim press. A provision to later shift the dies can be planned. Loose-fitting, die-mounting holes or slots are machined into the die-mounting struts, and roll-pin holes are added to assist in locating the final permanent die position. Initially, only two of the four pins are drilled into the mounting plate to provide the calculated home-position design location. If the dies need to be shifted to adjust for shrinkage, the other two unused locator roll pins can then be drilled, and the loosely fitted mounting bolts secured.

## Spare Dies

At the time of purchase of forged dies, it is prudent to buy spares and to ensure that the height and perimeter dimensions allow the interchange of dies in case of damage. (One common die accident is striking a loose screwdriver or bolt.) The die maker can refinish damaged dies, but it usually means shipping all the dies and the plate back for regrinding and sharpening. When selecting a die maker, it is important to be certain that the heat-treating facility provides consistent results and all dies are checked for uniform hardness. All new dies look keen and sharp. However, unless the heat treatment is done according to specification, they will dull quickly.

Manually fabricated forged dies can be supplied to a tolerance of ±0.015 in. (±0.38 mm). The trim-die perimeter can be machined to a tighter tolerance at increased cost.

## Machined Tool-steel Dies

Machined tool-steel dies are milled from 0.50–1.00-in. (12.7–25.4-mm) thick, bar-tool steel stock and the knife-like shape is produced on a vertical milling machine. The rough die blank is heat treated, ground to height, and the edge is hand-sharpened. Depending on the tool steel used, the hardness and toughness characteristics must be sufficient to avoid brittleness, which can cause breakdown of the cutting edge during service. Individual dies may be mounted on support blocks to provide depth to clear the thermoformed parts. The complete assembly, in turn, is fastened to a base plate (see Figure 10-7).

The tool-steel die can be machined to the closest tolerance (±0.001 in. [±0.03 mm]) of all knife-like dies. Machined dies re-

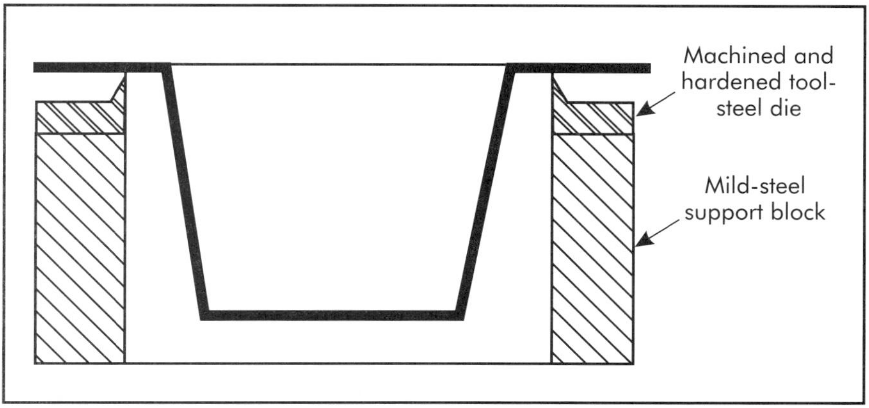


*Figure 10-7. Machined and hardened knife-like die backed by a low-cost support block.*

quire expensive raw materials, extensive machining, careful heat treatment, and manual sharpening. They have the highest cost among knife-like dies.

## Punches

Individual standardized punches can be used in conjunction with all of the different knife-like dies to trim out round holes, standard shapes, and forms, such as hang holes, for supermarket blister packaging (see Figure 10-8). When specifying punches, a major problem is how to handle loose scrap segments.

The scrap segment can be pushed back into the body of the trimmed part by inserting stripper-sponge rubber or a spring-action mechanism in the punch center (see Figure 10-9). The scrap segment will then be carried out with the trimmed parts and away from the press for later removal.

## Methods of Scrap Removal

There are several methods of scrap removal: manual detachment, semi-automatic removal, self-stripping, and manual punching.

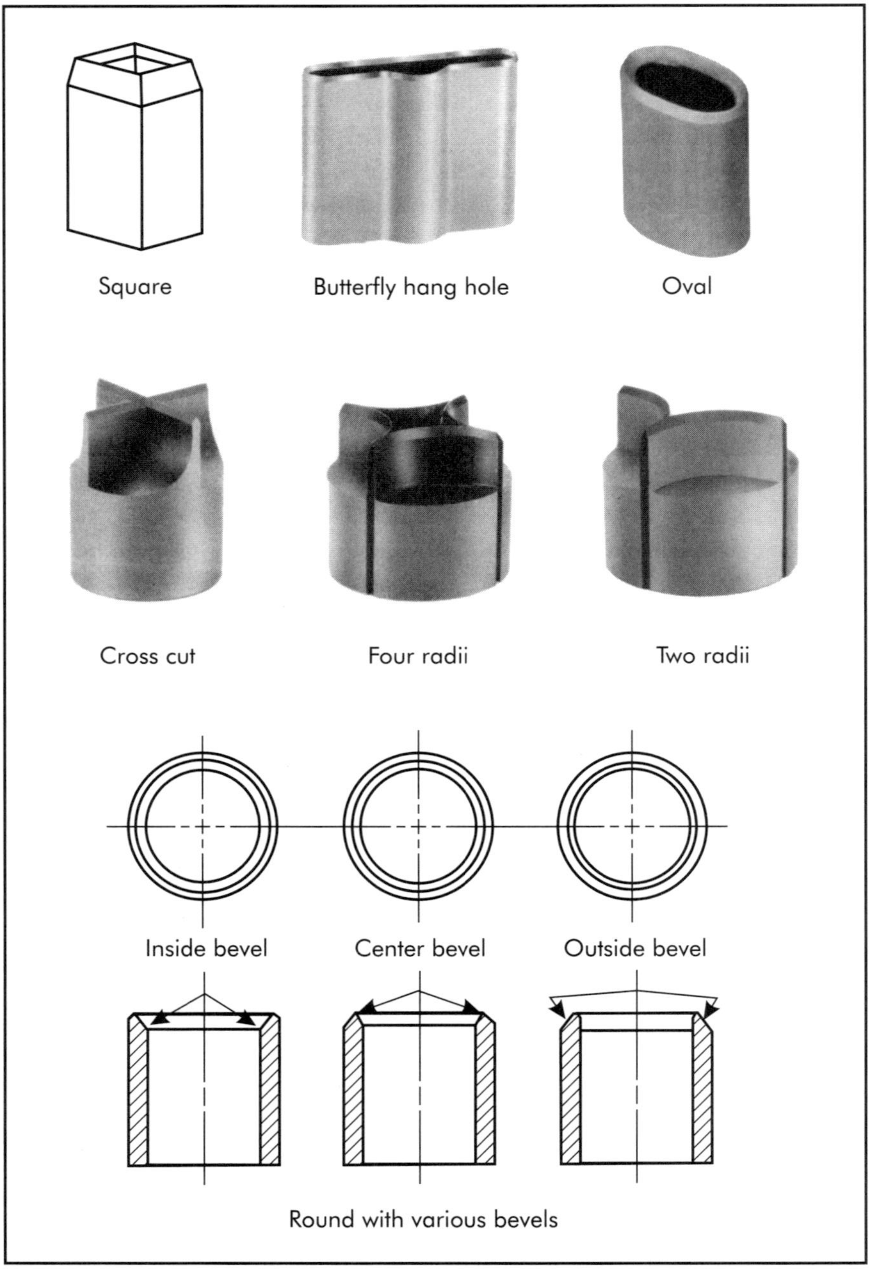


*Figure 10-8. Typical standardized steel-rule punches. (Courtesy Frank Wagner & Sons and J.F. Helmold Corp.)*

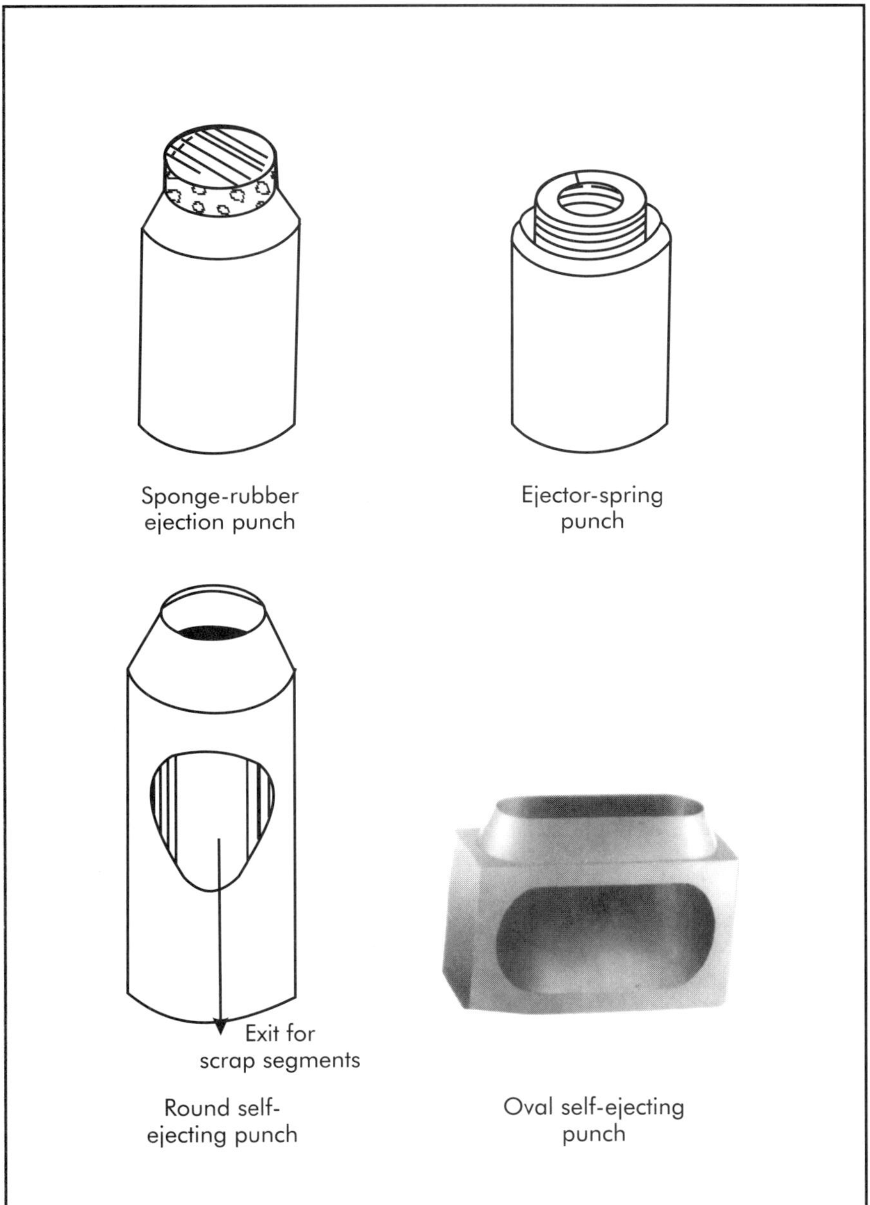


*Figure 10-9. Methods of ejecting scrap from punches. (Courtesy Frank Wagner & Sons)*

### *Manual Detachment*

A customer may accept thermoformed parts with the punched segment still inserted in the holes. These parts can be detached by simply pushing them out. The customer could receive a financial incentive for this labor.

### *Semi-automatic Removal*

Some of the in-line thermoforming machines have an automatic part-stacking station, which could be set up with the appropriate tooling to push out the die-cut scrap. This method may then require manual removal of finished trimmed parts from the web because most thermoforming machines come equipped with only one automatic part-stacking station.

### *Self-stripping*

The die could be designed with self-ejecting punches on the bottom platen of the trim station (see Figure 10-9). The scrap segments, when trimmed, would remain in the entrance of the punch and the next part trimmed would push the scrap deeper into the punch. The interior of the self-stripping punch will gradually fill with scrap that can exit from an outlet cut in the side of the punch. The scrap can then be removed from the die-mounting plate. Self-stripping punches are notorious for jamming with scrap plastic, bursting, and thus disrupting continuous die-cutting production. When the punches do eject scrap properly, loose scrap pieces may migrate into the machine mechanism or may be attracted to the web by the static electric charge generated by indexing plastic webs. A remedy for this problem must be sought. Compressed air or a vacuum system has been used to gather much of the loose plastic scrap out of harm's way under favorable conditions. However, either is not a perfect solution. Positive methods exist to remove punched scrap from thermoformed parts, but they are labor intensive or costly.

## *Manual Punching*

Each formed and trimmed part in a single-cavity punch and die can be manually punched using a pneumatic or kick press. The punch will drive the scrap segment into the die and gravity will cause it to fall into a collection container. The punch and die is mounted in a die set to maintain correct tool alignment and cutting clearance. The production rate using this manual method is slow and expensive, although the tooling cost is much lower than building a multicavity punch and die.

A similar punching scheme can be used to punch out scrap segments from each thermoformed shot, using a multicavity punch and die to remove segments with one stroke of the press. This system calls for a big press bed to accommodate a large, relatively expensive die. It is imperative that perimeter-trimmed cavities be held in place by weak tabs. This is due to the manual handling, which may cause the trimmed cavities to fall apart before reaching the multicavity die. If cavities in the shot are not pretrimmed, a later, manual-die-cutting operation to separate individual cavities from the shot is required.

## Hybrid Dies

A hybrid die to fit an in-line trim press can be designed for very long continuous runs of thermoformed parts, which require clean punching of segments with minimal manual labor. Hybrid multicavity dies are a combination of a punch and die for easy scrap-segment removal and forged dies to cut the perimeter of thermoformed parts (see Figures 10-10 and 10-11). The punch and dies require a four-post die set to maintain punch-and-die alignment and a scrap removal system for loose segments.

In-line trim presses with a die set are located within a set of parallel sheet-indexing rails traversing through the press platens. This setup requires that die-set posts and bushings be placed on the exterior of these rails. The posts must be easily removed to allow the bottom section of the hybrid die set to be installed in the press. The lower platen of an in-line trim press does not contain

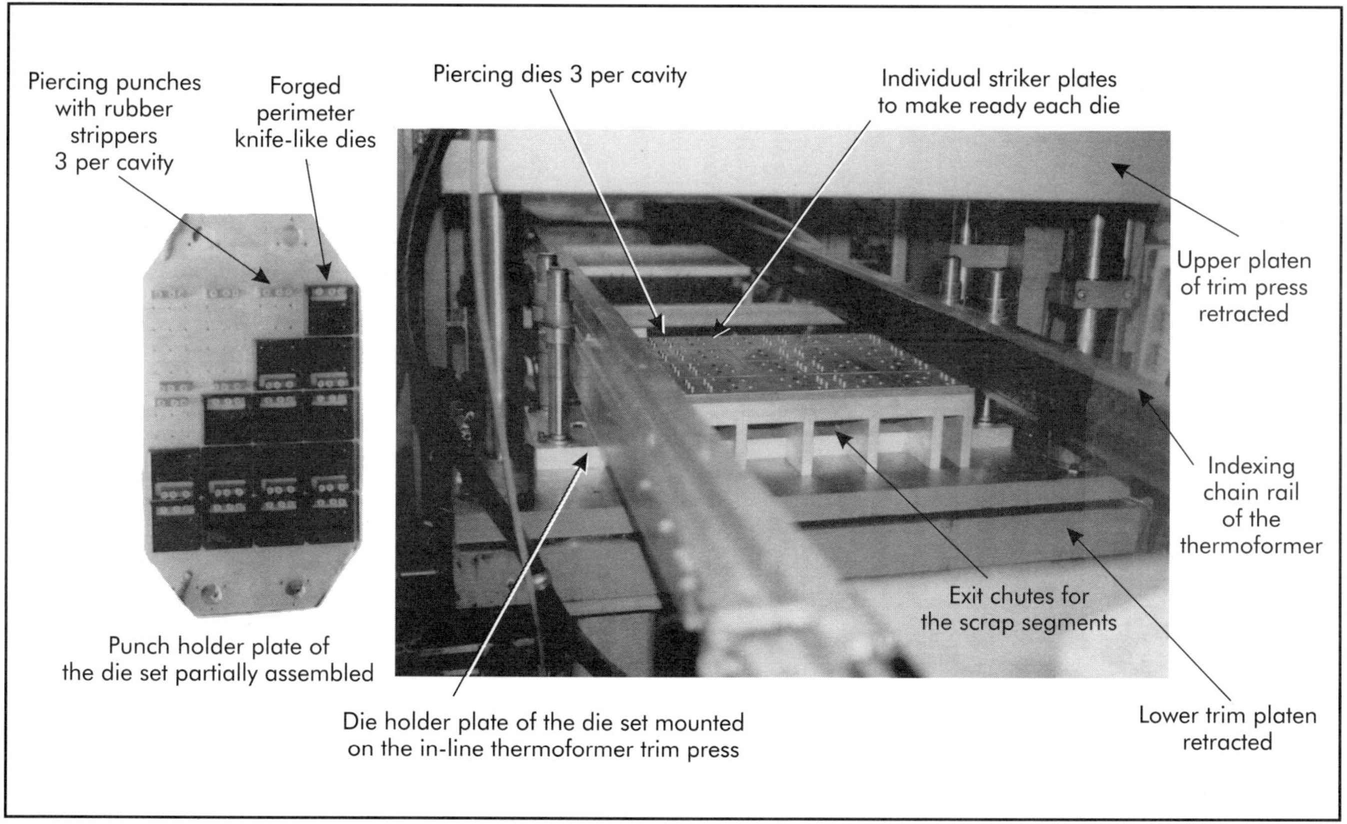


*Figure 10-10. Hybrid segment punch and die with forged perimeter knife-like dies.*

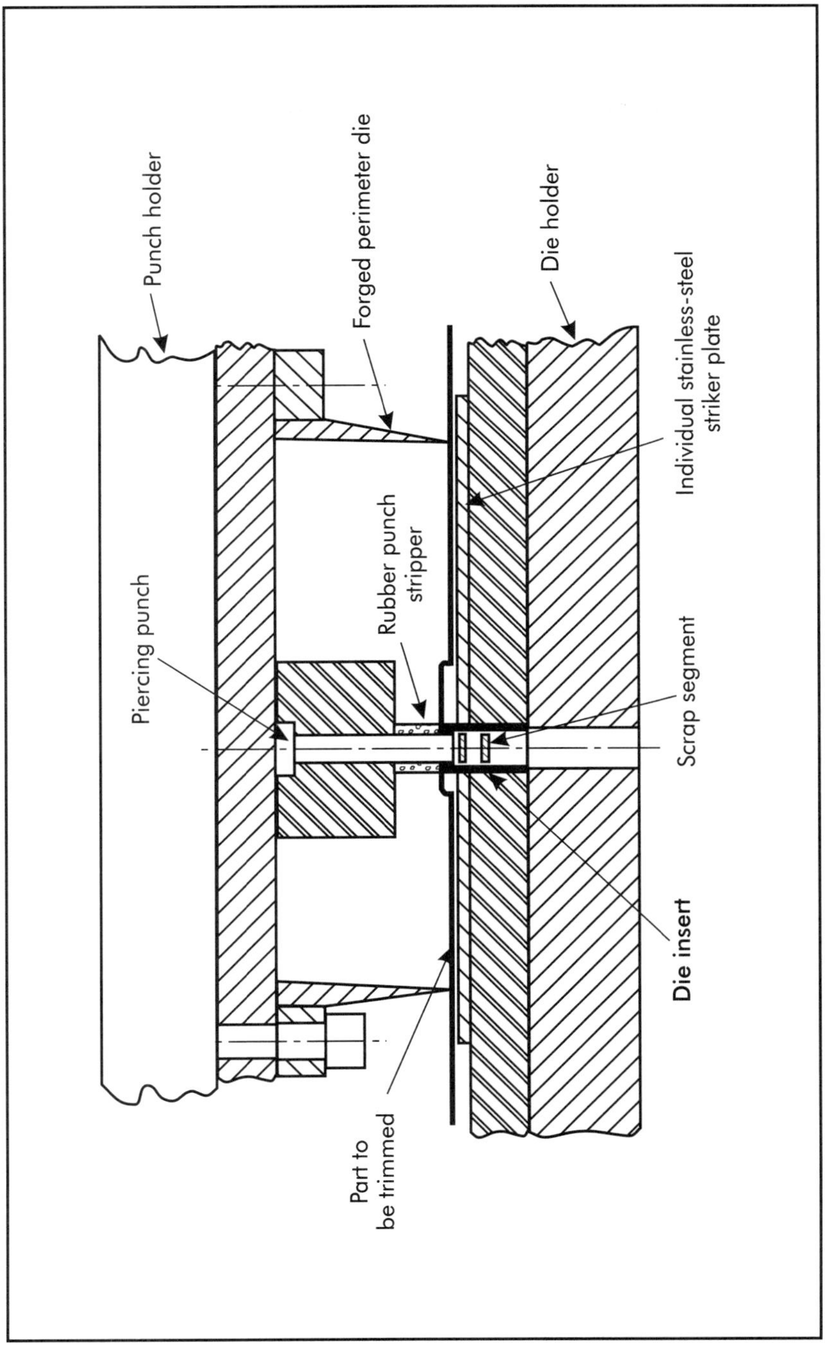


*Figure 10-11. Hybrid die cross section.*

an opening for loose scrap segments to fall easily through the platens. The hybrid die should include a system, such as a low-profile conveyor belt or gravity scrap chutes, to move punched scrap away from the die area.

Make-ready striker plates for each cavity assist with clean trimming of the part perimeter by hybrid forged dies. Plastic shrinkage allowance for location of the punch and dies is critical since they are immovable, unlike forged-perimeter dies, which can be shifted if necessary. Mold design should consider movement of mold cavities, if needed, to line up with the punches to achieve registration. Each punch needs a stripper device to press down the formed part against the die as the punch enters the die. The stripper device can be a donut-shaped rubber or polyurethane section slipped onto the punch and held by friction. The leading edge of the stripper material extends past the punch tip and compresses when cutting, providing stripping action when the die opens.

Most commercial die sets are built of steel and the plates can be quite heavy, particularly with necessary buildup plates. They may exceed the specified maximum tooling weight for the trim press. More importantly, plate weight can surpass the handling capacity of the thermoforming shop setup crew. Instead of steel, a hybrid die assembly of aluminum jig plate is one-third the weight and has good stability. All the punches and dies used are fully hardened and press-fitted into aluminum retaining plates. The removable die-set pins and bushings are aligned by hardened and ground steel bushings mounted in the die set.

Punching out scrap segments appears to be a minor trimming problem, but it always involves nuisance and cost. The die maker and the customer should decide what available options suit their particular project.

## Heat-assisted Die Cutting

Heat-assisted die cutting reduces the force needed using a knife-like die. Properly transmitted heat lowers the shear strength of plastic when the temperature is raised to softening temperature. Some thick plastic sheets (high-impact styrene, acrylics, etc.) may shatter, causing rough-trimmed finished edges and cracks when an attempt is made to drive the die knife at room temperature. These cracks can extend into the finished parts. This problem becomes extreme as plastic thickness increases; a 0.25 in. (6.4 mm) sheet when cut cold emits sounds like a shotgun as the knife-like die breaks its way through the plastic sheet as if it were glass.

Most in-line trim presses have a low-pressure closing stroke that can be timed to the heat-assisted sequence before high-pressure cutting is activated. A uniform temperature must be available during the process, and a stable heat reservoir can be obtained from heat buildup in the heavy striker plate when it is electrically heated. The temperature is gradually raised until it is below the temperature at which hot plastic sticks to the striker plate. The press platens are adjusted during the low-pressure press extension cycle so that the knife does not cut through, yet is in complete contact with the plastic. A heated striker plate only heats the spot where the knife impinges on the plate and does not disturb or distort the rest of the formed part. During the trim cycle, a layer of air insulates the rest of the part from the heated plate during the short time of the trim cycle.

Some thermoforming machinery manufacturers recommend heating the die rather than the striker plate. It is questionable whether sufficient heat can be driven from the base of the thin steel-rule blade to its cutting edge to provide a consistent temperature for die cutting. The standard steel rule is imbedded in a wooden die board, which tends to shrink and crack as moisture is driven from the wood during heating. There may be some justification for this technique if a steel-rule die is constructed using an aluminum die-board base and die allowance is calculated for the metal base to expand during heating.

Heat-assisted die cutting does not eliminate the need to make-ready dies that are not cutting properly. All the knife blades must be in contact with the striker plate to achieve uniform trimming.

## Trimming on Two Planes Using Knife-like Dies

Trimming on two planes using knife-like dies demands a great deal of make-ready skill from the thermoforming setup crew (see Figure 10-12). When die cutting is in one plane only, each of the

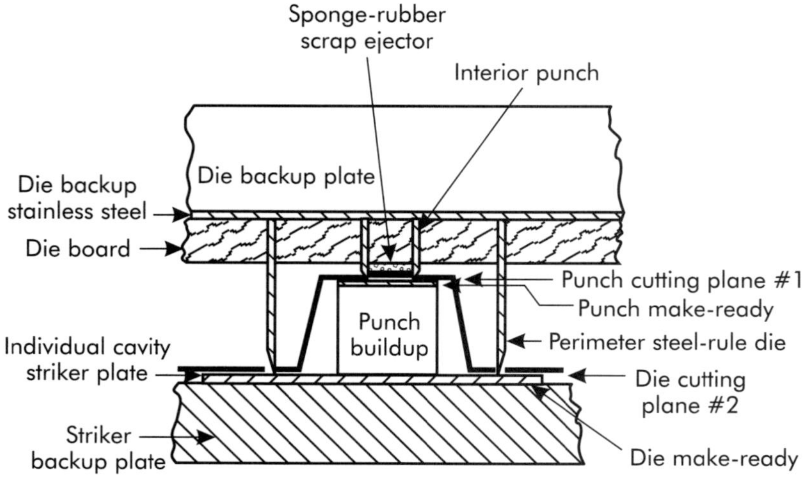


*Figure 10-12. Trimming a part on two planes using a knife-like die.*

cavities of the die must be made ready with reference to each other for the whole shot to be cleanly trimmed. Die cutting on two planes requires that two separate dies trim on individual planes and not obstruct each other. If one of the dies cuts perfectly in one plane after addition of shims, these shims may affect the ability of the second die to cut cleanly in the second plane. The second die may then need make-ready shims for its striker plate to cut cleanly, causing a disturbance of the original make-ready for the first die.

When designing a two-plane die, it will be useful if the die-setup technicians provide individual-cavity striker plates for the perimeter die cavities and prefabricated 0.001–0.002-in. (0.03–0.05-mm) thick make-ready shims for the interior die. These thin shims are applied slowly. Both dies are continually tested to make sure the new make-ready does not change the cutting pattern of the first die, and that both planes cut simultaneously. All of the plates used for the dies and their buildups must be flat, burr-free, and of precision thickness to avoid creating additional obstacles.

## Trimming Above and Below the Sheet Line

Die cutting of parts formed above and below the sheet line requires furnishing an additional die component—a match plate. Openings must be provided within both the die and its match-plate assembly (consisting of a striker plate and its buildup) to clear thermoformed part protrusions (see Figure 10-13). The die and its match plate are made to register by means of lineup holes for removable alignment pins in each of the tool halves. The match-plate part-clearance opening should be chosen from the thermoformed component side that provides the widest remaining striker-plate land for the die blade to cut against. A knife-like die cannot have its blade strike inside the open edge of a match plate because the die will become damaged and the cutting erratic. Most in-line die-cutting presses have both platens movable; and the top platen often has a greater stroke than the lower platen. Care must be exercised that sufficient part clearance is available from each platen chosen for the match-plate-and-die combination.

Trimming parts formed above and below the sheet line leaves no support under individual parts after the trim. The parts are suspended without support in the web. Only uncut tabs (nicks)

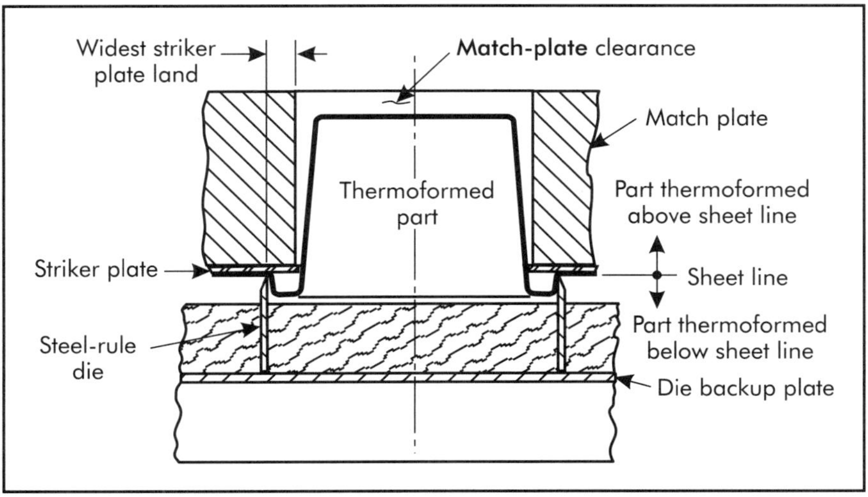


*Figure 10-13. Trimming a part formed above and below the sheet line.*

keep them from breaking free when both press platens are opened. Stripper rubber helps prevent parts from "hanging up" or sticking in the die. These hang-ups often cause attachment tabs to break when the platens open. To avoid disturbing trimmed parts, the machine operator should adjust the cycle so the die platen will retract first (allowing the part to strip from the die) before opening the match-plate platen.

Combining the processes of die cutting and lamination (laminating a backing sheet to a thermoformed part) provides two operations in one stroke of the trim press. This type of part design increases the rigidity of a package constructed of a floppy thermoformed part laminated to a backing sheet of cardboard or plastic (see Figure 10-14). Production can take place on a continuous in-line thermoformer or on manual off-line die cutters. A standard, steel-rule die is modified for this service with spring-activated pressure pads fitted inside the die close against the steel rule.

The backing sheet is coated with a suitable adhesive and placed on the flange of a downward-facing thermoformed shot before it enters the die-cutting press. When the die-cutting platens are closed, the die pressure pads press against the flange area of the part, laminating the components at the same instant the knife trims the cavity. Care should be exercised in choosing the adhesive; it should remain tacky until laminated and then set quickly. When the package is sealed, considerable liquid solvent can re-

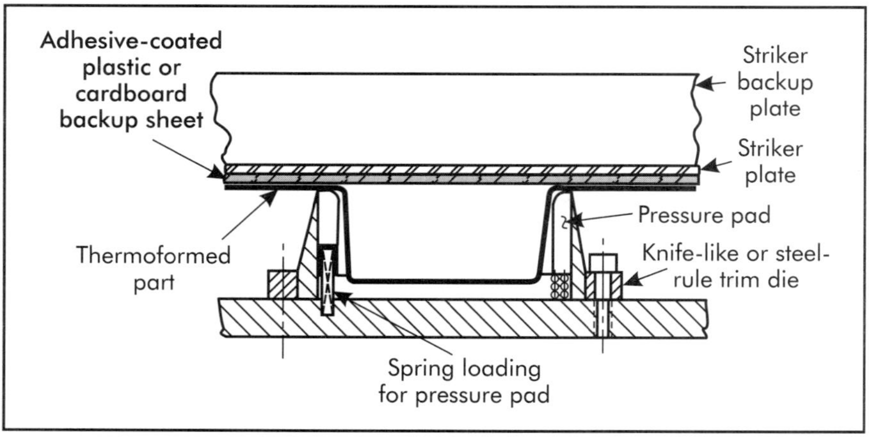


*Figure 10-14. Knife-like die for simultaneous adhesive lamination and trimming.*

main entrapped inside the laminate and create a source of future damage. Therefore, the adhesive should be selected so that the plastic is not affected.

A similar die configuration can be used to heat-seal a heat-activated, adhesive-coated cardboard backing to a thermoformed component. The platen opposite the die requires a temperature-controlled platen to activate the coated, heat-seal board when the platens are closed during a timed interval. The pressure pads should have a silicone, sponge-rubber face similar to blister-seal tooling (see Figure 10-15). When heat-sealing the laminate, a minimum 0.25 in. (6.4 mm) flange is recommended for package strength and to allow the silicone rubber sponge to properly adhere to its metal backup pad.

## Choosing a Knife-like Die

A thermoformed part has a high-quality trimmed edge if the perimeter of the part feels smooth and clean to the finger. When the knife edge is dull, it will crush and break through the plastic, leaving a lumpy, uneven edge on the trimmed wall. A dull die has a flat shiny face visible on its cutting edge. This flat area crushes a thin segment of plastic, which results in a residue called *angel hair* adhering to the finished part. Angel hair is very difficult to remove because it is often held to the finished part by static electricity.

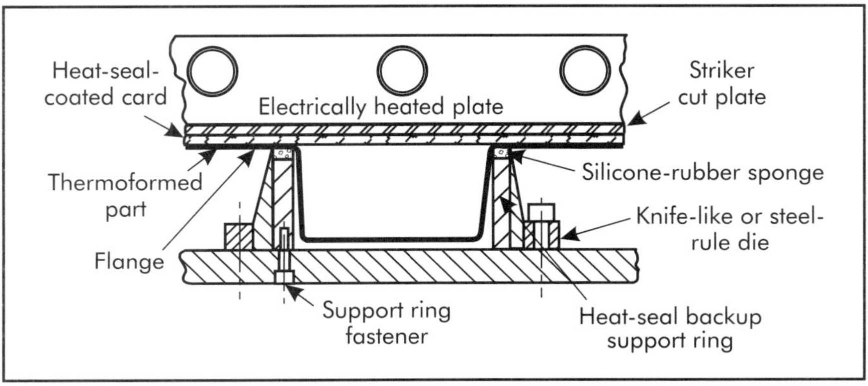


*Figure 10-15. Tooling for a combination heat-sealing and trim die.*

Medical, food product, and single-use catering product users consider angel hair to be a contaminant; other less-critical industries look on it as an avoidable nuisance that a sharp knife edge can minimize. Table 10-1 summarizes the characteristics of knife-like trim dies.

Economics play the most important role in knife-like die selection with a diminishing importance paid to quality of the trim, die life, ability to shift die cavities, and die delivery time. The hardness of the cutting edge helps to determine the life of the cutting edge. However, the user is responsible for the knowledge and skill to properly make-ready a new die. The cutting edge of any knife-like die can be destroyed with the first stroke of the trim press on an improper make-ready. Table 10-2 shows the advantages and limitations of four types of knife-like trim dies.

## Preliminary Layout of Knife-like Dies

Data needed to lay out a die comes directly from the mold design, specified plastic resin, customer requirements, and thermoforming machinery trim-press data.

The die layout can proceed smoothly after the following questions have been answered:

1. Will the parts be formed above or below the sheet line or will they be both below and above?
2. What is the resin shrinkage coefficient (see Chapter 8, Table 8-1) and will trimming be in-line (half-shrinkage) or off-line die cutting (full shrinkage)?
3. What type of knife-like die best suits the customer's needs? What is the quality and quantity of parts, delivery schedule, and proposed die budget?
4. Are there any existing die-buildup plates available for use with this die, or are new buildup plates to be supplied?
5. Does the in-line trim-press data show whether upper and lower platen stokes will fit the die requirements?
6. Special die specifications that might be requested by the customer:

Table 10-1. Knife-like trim-die characteristics

| Type of Die | Ratio of Cost When Compared to Steel-rule Dies | Typical Delivery (Days) | Quality of Trim | Die Hardness ($R_C$) | Average Life (Strikes Until Sharpening or Replacement) |
|---|---|---|---|---|---|
| Steel-rule die | 1:1 | 1–5 | Fair | 35 | 1,000–5,000 |
| Knife-like forged die | 3:1–5:1 | 10–20 | Good | 45–56 | 5,000–10,000* |
| Machined tool-steel die | 5:1–8:1 | 15–30 | Best** | 45–56 | 10,000–15,000* |
| Hybrid forged die and punch and dies | 6:1–12:1 | 60–90 | Very best | Punches 59–61 | Punches* 20,000–50,000 |

* Can be resharpened by a knowledgeable customer
** Most accurate die ±0.001 in. (±0.03 mm)

Notes:
1. Cost and delivery are directly proportional to the complexity of the die.
2. Average die life is dependent on many factors; for example, plastic material characteristics, ability to make-ready the die, trim-press platen parallelism, etc.

Table 10-2. Knife-like die comparison

| Type of Die | Advantages | Limitations |
|---|---|---|
| Steel-rule dies | Lowest cost, rapid delivery, and many die suppliers available. Easily re-ruled. | Dies cannot be sharpened, shorter die life, dulls rapidly, quality of trimmed parts deteriorates rapidly, center-to-center errors cannot be corrected. |
| Knife-like forged dies | Trim edges stand up 2-3 times as long as steel-rule dies, can be resharpened, center-to-center and die errors can be corrected. | Few die suppliers available; delivery time is longer than for steel-rule dies. |
| Machined tool-steel dies | Longest production die life, can be re-sharpened, most precise die, center-to center errors can be corrected, and local tool shops can fabricate. | Most expensive knife-like die type, long delivery time. |
| Hybrid dies—forged die with punch and die | Built in a die set. Clean trim out of punch segments. | Most expensive, longest delivery time for this type of die. |

A. Will this be heat-assisted die cutting? Is there an existing heat-assisted striker plate assembly or it is to be supplied?
B. Are there laminating or heat-sealing requirements?
C. If punches are needed, how are the scrap segments to be handled?
D. Method of assuring accuracy of registration of shot to the die:

a) Are the forged die cavities to be individually relocated by the customer as needed?

b) Is a standard steel-rule, one-piece die to be floated or fixed to the trim-press platen?
c) Are individual die cavities to be floated?
d) Are pilot locators to be provided?

E. Is sponge or mechanical part stripping to be specified?

7. If a match plate is required, will it be attached to an existing buildup or is a complete buildup necessary?
8. What type of striker plate is needed—stainless steel or hardened-steel plate?
9. On a deep-cavity die, is the pedestal buildup depth beyond the back of the die to be metal or constructed of die board?

## Press Tonnage

Every model of trim press has a tonnage output rating that can be used to calculate the maximum length of knife blade that may be incorporated into a die. All the following design data is empirical, but the information provides a conservative guide to the design of many dies.

- Straight lengths of knife require 100–300 lb (17.5–52.6 N) of force per running in. (mm) of die to cut at room temperature.
- Small radii of 0.06–0.19 in. (1.5–4.8 mm) offer more cutting resistance than larger radii. Their circumferential length should be increased to two-to-four times that of an equivalent straight knife length.
- The perimeter of all punches with rubber internal knockouts absorbs a great deal of press tonnage—two times their circumference length.
- Plastic resins vary in resistance to knife trimming and are listed by increasing difficulty:
  high-impact styrene (HIPS) (least difficult); polyvinyl chloride (PVC); polyolefins; acrylonitrile butadiene styrene (ABS) and engineering plastics; and polyethylene terephthalate glycol (PETG) (most difficult).
- The knife blades gradually become dull during production, which may cause the cutting force to double before the blade is resharpened or replaced.

- Heat-assisted trimming can reduce by half the required cutting force of an ambient-temperature die.

To find the maximum allowable blade length required for a given press tonnage, consider the following example.

Maximum trim-press cutting force—200,000 lbf (890 kN) available
Material to be trimmed—HIPS or PETG
Trim perimeter—rectangle 5 × 7 in. (127 × 177.8 mm)
Corner radius—0.50 in. (12.7 mm), four required
Punches—0.50 in. (12.7 mm) diameter, three required with internal knockouts

Determine the number of in. (mm) of straight rule in each cavity by using Table 10-3.

HIPS—Estimated cutting force requirement = 100 lbf (17.5 N) per in. (mm) of length
PETG—Estimated cutting force requirement = 300 lbf (52.6 N) per in. (mm) of length

$$N_c = \frac{F_F}{C_F P_C} \tag{10-1}$$

where:

$N_C$ = maximum number of cavities to be trimmed
$F_F$ = trim-press cutting force, lbf (N)

Table 10-3. Projected lengths

| Cavity Component | Calculated Length | Difficulty Factor* | Knife Length in. (mm) |
|---|---|---|---|
| Rectangle | 2 (4 + 6)** | ×1 | 20 (508 mm) |
| Punches | 3 required × π × .5 | × 2 (punches)<br>× 2 (rubber) = 4 | 18.80<br>(477.5 mm) |
| Corner radius | $\frac{4 \text{ required} \times \pi \times 1}{4 \text{ corners}}$ | × 2 | 6.28<br>(159.5 mm) |
| Total projected lengths of rule for each cavity = 45.08 in. (1,145.0 mm) | | | |

* Difficulty factor—judgment call based on radius size compared to straight length.
** Straight length less end radii.

$C_F$ = plastic resin cutting force, lbf (N)
$P_C$ = total projected rule length for each cavity, in. (mm)

$$\text{HIPS} = \frac{200{,}000}{100 \times 45.08} = 44 \text{ possible cavities}$$

$$\text{PETG} = \frac{200{,}000}{300 \times 45.08} = 15 \text{ possible cavities}$$

For a dull die, assume doubling the trim force (very conservative) to compensate for long production.

$$\text{HIPS} = \frac{44 \text{ cavities}}{2} = 22 \text{ cavities}$$

$$\text{PETG} = \frac{15 \text{ cavities}}{2} = 7\text{-}1/2 \text{ (round out to 8 cavities)}$$

The average thermoforming machine could profitably run 22 HIPS cavities per cycle. Eight PETG cavities per shot is an uneconomical quantity. Tooling should be augmented with heat assistance to double the output to 16 PETG cavities per cycle.

Aligning the center of pressure of die-cavity perimeters to the centerline of the press distributes the trim-press forces evenly to each cavity. Where the press and the die centerline coincide, the center for a die cavity can be found if it has a regular shape, such as a circle or a rectangle. This is not true of irregular shapes such as triangles, semi-hemispheres, etc., as their center of pressure (CP) is equal to the center of gravity (CG) of the lines of their perimeters. Some geometric shapes with their perimeter lines and center of gravity are illustrated in Figure 10-16. The lines of the perimeter and its center of gravity are not the same as the center of gravity of the shape's area.

Example: Find center of pressure (CP) of a die by using Equation 10-2. Refer to Figure 10-17 for data.

$H_X$ = distance from $X$–$X$ axis to CP of the die (10-2)

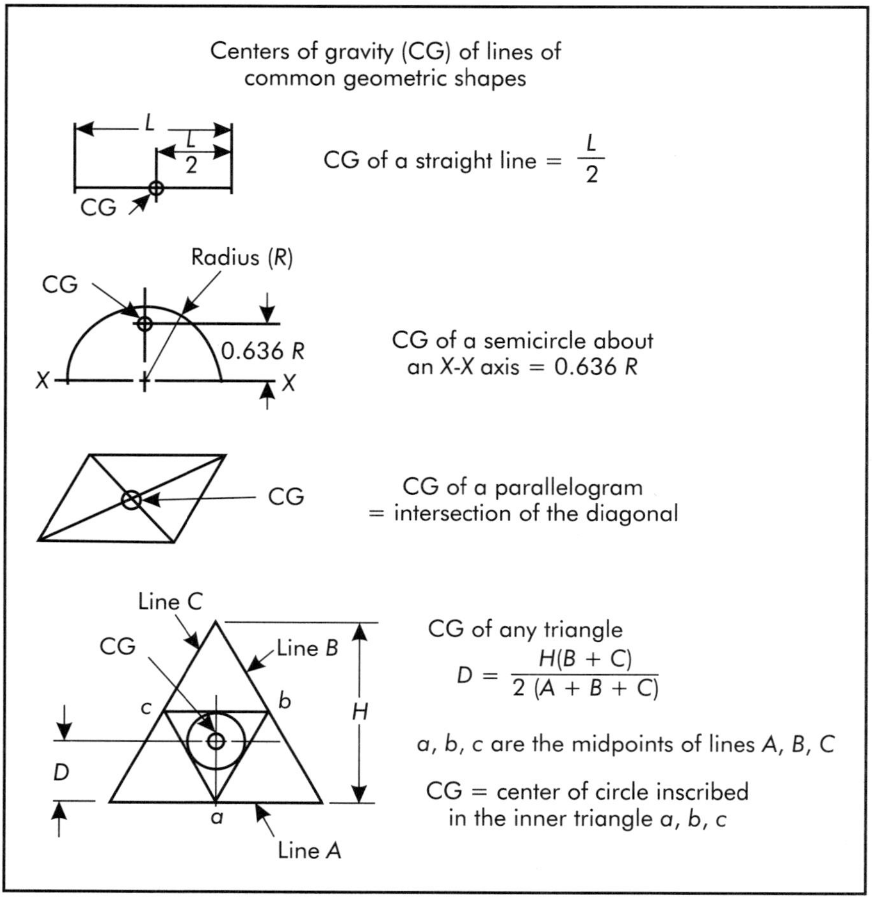


*Figure 10-16. Perimeter lines and centers of gravity for some geometric shapes.*

$$X_1 = X_2 = \frac{\sqrt{4^2 - 3^2}}{2} = 1.322 \text{ in.}$$

$$\text{semicircle CG} = 0.636\,R = 0.636(3) = 1.908 \text{ in.}$$

$$X_3 = 1.908 + 2.645 = 4.553 \text{ in.}$$

$$L_1 = L_2 = 4 \text{ in.}$$

$$L_3 = \frac{\pi D}{2} = \frac{\pi \times 6}{2} = 9.42 \text{ in.}$$

$$H_x = \frac{L_1 X_1 + L_2 X_2 + L_3 X_3}{L_1 + L_2 + L_3} = \frac{2(4 \times 1.322) + (9.42 \times 4.553)}{2(4) + 9.42} = \frac{53.48}{17.42}$$

$$H_x = 3.07 \text{ in. CP of die}$$

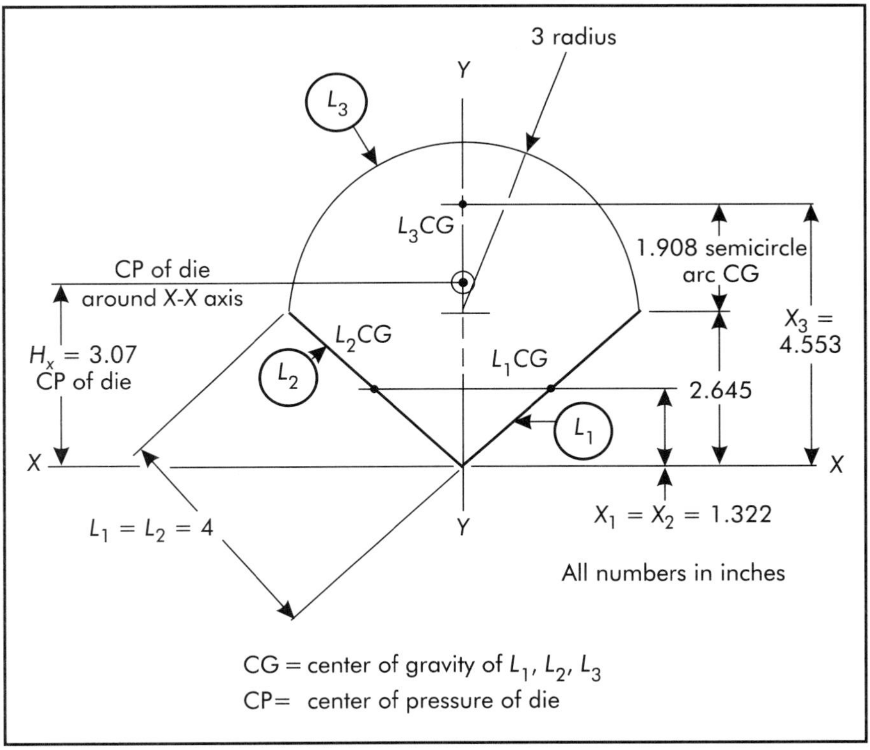


*Figure 10-17. Center of pressure (CP) for a simple knife-like die.*

## Layout of a Steel-rule Die

Mold layout dimensions are used as a basis for die layout. Each centerline dimension should be identified as to whether it is for the mold or for the die. The die dimensions are reduced by the styrene (HIPS) shrinkage coefficient for in-line thermoformers of 0.0025 in./in. (0.0025 mm/mm) as shown in Figure 10-18. The die board minimum requirement of 0.25 in. (6.4 mm) of wood between die cavities is necessary to maintain the structural integrity of the board. The mold cavity center-to-center distance of 5.74 in. – (5.74 × 0.0025 sheet shrinkage) = 5.726 in. (145.44 mm) die center-to-center distance. This is the starting point of the die layout.

The die board overall dimensions are smaller than the mold base to provide clearance if the board must pass through the indexing rails of the machine. Die cavities have sufficient board-edge support (0.74 in. [18.8 mm]) and the blade extension beyond the board thickness (0.31 in. [7.9 mm]) will provide a sturdy backing for the steel rule.

The part depth of 3 in. (76.2 mm) requires an aluminum jig-plate buildup (2.25 in. [57.2 mm]) and 0.06-in. (1.5 mm) stainless sheet die-backup plate to supplement the die. A hole through the die board and its buildup to clear the thermoformed part completes this assembly. Mounting holes or other means of securing this die assembly to the trim press should be provided. If this rather heavy assembly is to be floated to improve registration, locator pilots and a limited movement system needs to be devised.

A die-dimensional layout similar to Figure 10-18 is required if this cup-like part is formed above and below the sheet line due to inclusion of a stiffener rim in its design (see Figure 10-13). The rim portion faces the steel-rule die and clears the face of the die board. The height of the cup extends up through the match-plate clearance hole, with trimming taking place at the striker plate. Lineup of the die to the match plate is accomplished by employing removable pins placed in holes drilled on the *X* or *Y* centerlines.

## Die User and Die Builder Responsibilities

Trimming thermoformed parts using knife-like dies is always less costly and the dies can be delivered more rapidly than punch and dies. However, knife-like dies require the user to develop necessary make-ready and die-cutting skills to provide high-quality trimmed parts and extend the life of the die blades. In-line die

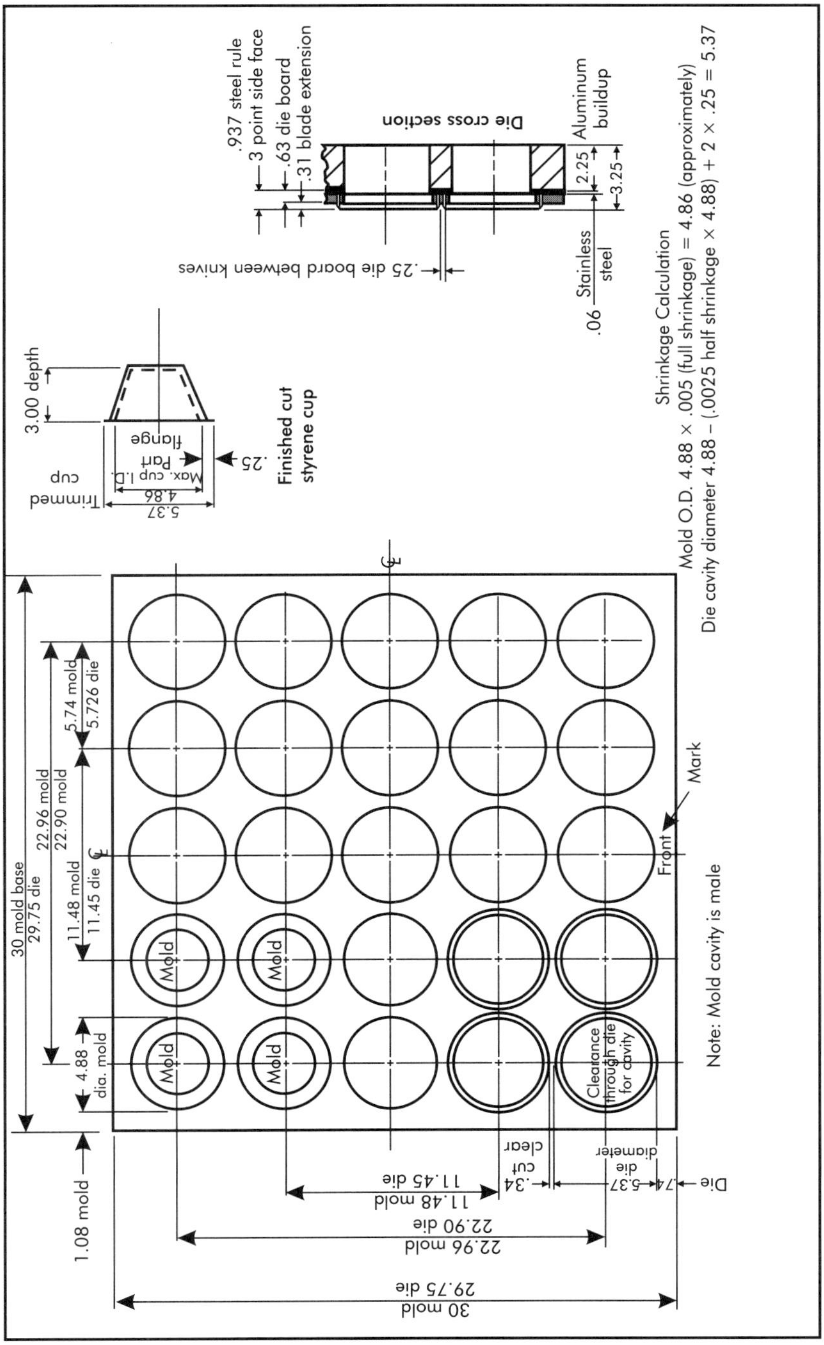


*Figure 10-18. Steel rule layout from mold dimensions, half-shrinkage.*

cutting on a thermoforming machine base requires making shot-shrinkage assumptions, which are not always precise when compared to off-line, fully cooled shot-shrinkage calculations.

Auxiliary devices—pilots, floating dies, heat-assisted trimming and stripping of parts—are tools for the designer to use to solve die problems before they occur in production. A decision on how to eliminate the scrap segments caused by punch trimming needs the input of the die user and the die maker. An informed customer should participate in deciding which types of knife-like dies will best serve the required production methods.

# 11

# Off-line Punch-and-die Trimming

Punch-and-die trimming provides the most economical and highest quality product when millions of thermoformed parts are required (see Figure 11-1). The dies stay sharp for many die strikes and can be resharpened on a tool-room surface grinder. Unlike knife-blade trimming, which requires careful make-ready to produce high-quality trimmed parts, punch-and-die tooling requires less training to master. *Make ready* is a system used to level all of the cutting blades of a knife-like die in one plane.

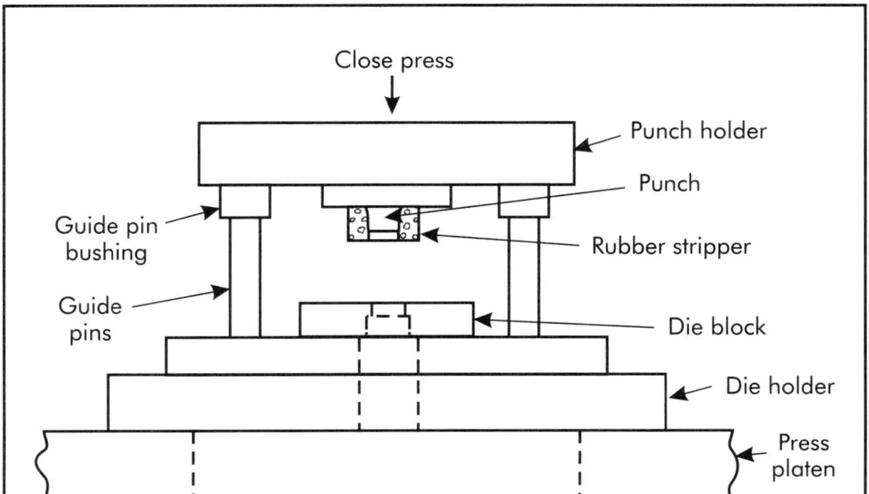


*Figure 11-1. Common components of a simple die (Nee 1998).*

The action of a punch and die provides true shearing action—superior to the splitting or crushing action of a knife blade—to produce a high-quality trimmed part (see Figure 11-2). The trimmed parts can automatically self-stack after being cut free from the web, allowing accurate counting and rapid packing of the completed units.

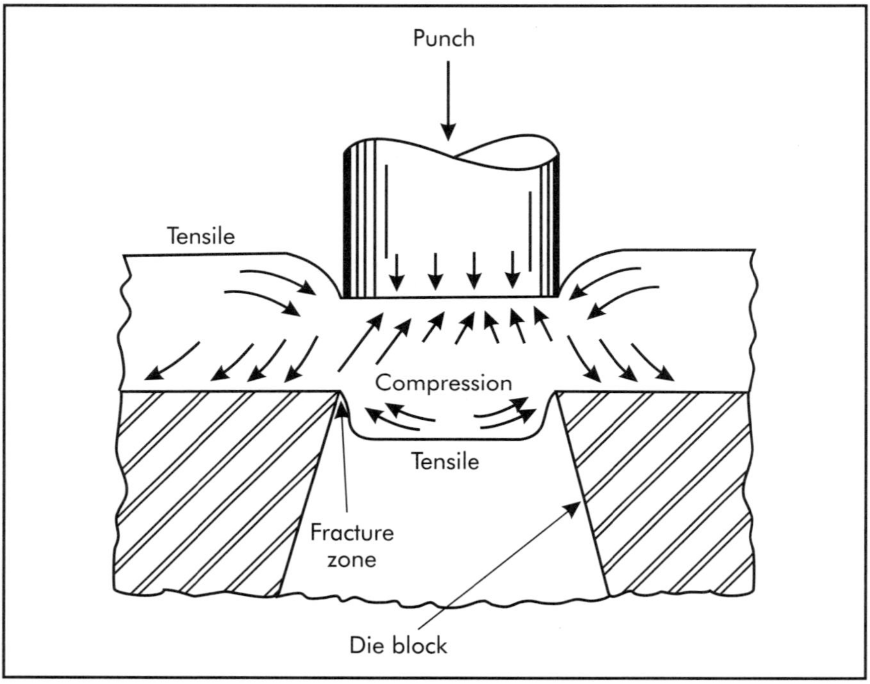


*Figure 11-2. Stresses caused by the action of the punch and die (Nee 1998).*

## Cost Justification

Most off-line trim presses are designed to allow trimming a row or two of a formed shot with each stroke of the press. The cost of a die consists of the fixed assembly (die set, buildup, scrap chopper, etc.) and the variable components (punches, dies, pilots, knockouts, and exit chutes). Each additional trim cavity increases the price of the variable components by an amount many times greater than the increase in cost of the fixed assembly. Therefore, the die designer must attempt to minimize the number of trim cavities in the die by a careful analysis of the press and its index system capabilities.

Punch-and-die initial purchase costs are higher and delivery time is longer than those for any other types of knife-like dies. Dividing the die cost by the number of parts to be trimmed provides the answer to the question of whether a punch and die is economical for a given project (see Table 11-1).

Table 11-1. Determining economy of purchase for a punch and die based on $25,000 price (1999)

| Number of Parts | Cost per Part, $ |
|---|---|
| 250,000 | 0.100 |
| 2,500,000 | 0.010 |
| 25,000,000 | 0.001 |

The cost per part in Table 11-1 suggests that production in volume of several million parts would be needed to amortize this tool for most practical purposes. Knife-like dies are often specified for part quantities that fall below the limits necessary to pay for a punch and die.

There are occasions when the advantages of trimming small quantities on a punch and die may outweigh the high initial cost.

- Parts whose trim perimeter may be in more than one plane or may need to be trimmed along a sidewall are best done on a punch and die.
- A properly maintained punch and die tends to produce far less angel hair than a knife-like die, which makes it appropriate for critical parts (medical and food packaging).
- A punch and die can cut very close to a thermoformed wall, leaving a miniscule flange far smaller than the tiniest flange a knife-like die produces.
- Parts cut on off-line trim presses generally allow the plastic web to cool to full shrinkage before being trimmed. A punch and die can help reduce shrinkage design error when used for cutting precision thermoformed parts.

## Punch and Die Design

The punch and die design should be based on the trimming needs, the trim press, and the maximum cutting force of the press.

### Trimming

Trimming can take place after one or more rows of parts are indexed into the cutting area of the die. Continuous thermoforming production at a maximum cyclic rate requires the trim press to be operated at speed sufficient to cut all rows of parts indexed into the cutting area of the die. The trim press and its indexer have a rated maximum revolutions per minute (rpm) specification for the flywheel that equals the number of trim strokes per minute. This rpm maximum specification varies depending on the length of the platen stroke, which can be manually adjusted to suit the height of the parts being processed. When the number of rows of parts thermoformed per minute exceeds the maximum allowable strokes/minute of the trim press, the number of die cavity rows should be increased or the thermoforming-machine output must be curtailed. Mold and die design should be coordinated so that the number of die rows can be divided into the number of rows of mold cavities by a whole number. This is done to avoid simultaneously trimming parts of two adjoining shots (for more information, see Chapter 8).

After the web has been trimmed of its usable components, the scrap skeleton is cut into strips suitable to be fed into a granulator. The scrap cutoff blade built into the die may be a straight-line guillotine shear or it may cut the skeleton into two or more parts for easier processing (see Figure 11-3). Cutting off skeleton scrap not only solves a material-handling problem, but also may prevent misregistration of the incoming web due to jamming of the scrap web. Cut skeletons can be removed by a conveyor belt from the die to be ground-up or dropped directly into the mouth of a low-profile granulator.

By incorporating two in-line die workstations (progressive die) in a single die set, the die can do more than blank out individual parts (see Figure 11-4). Where die space permits, the tool may

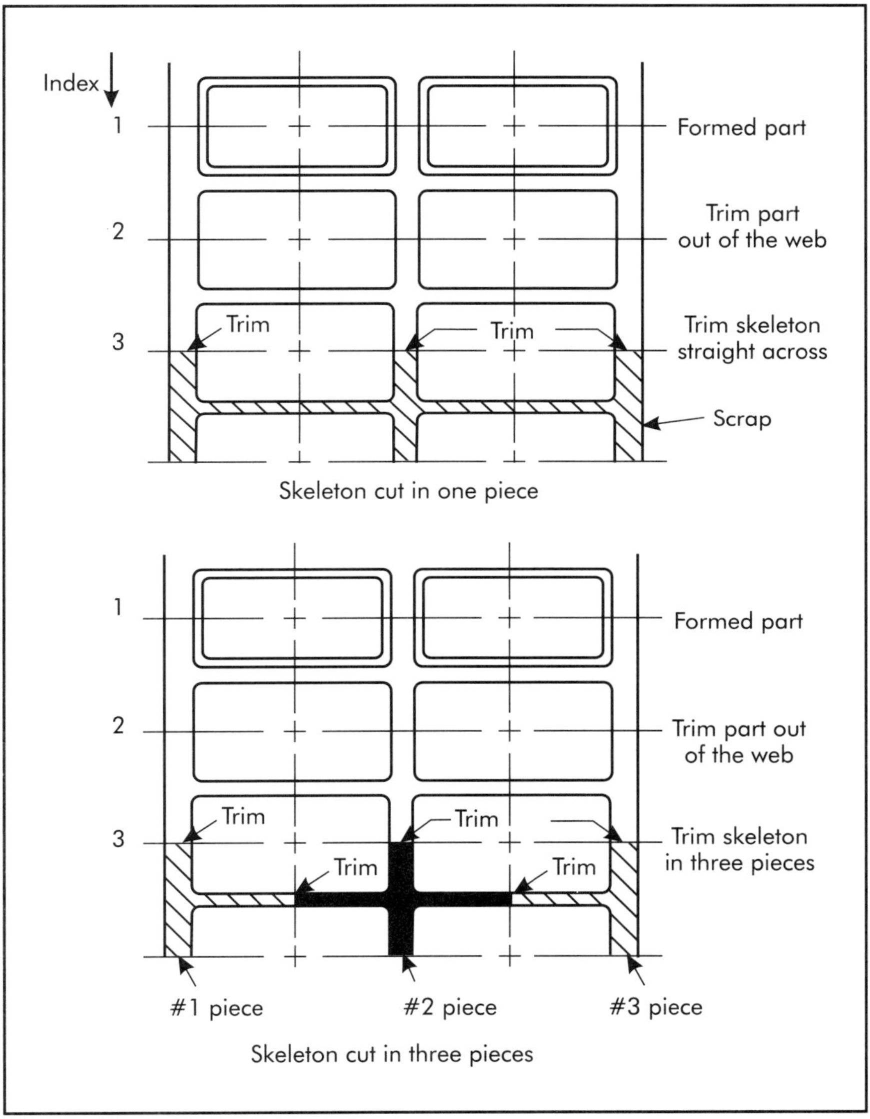


*Figure 11-3. Methods of cutting off scrap skeleton into one or more pieces.*

punch out segments in a first die station. Then, that row is indexed to the next station for final perimeter cutting. This technique saves the cost of building two dies and tying up an additional trim press. It works well for noncritical hole locations.

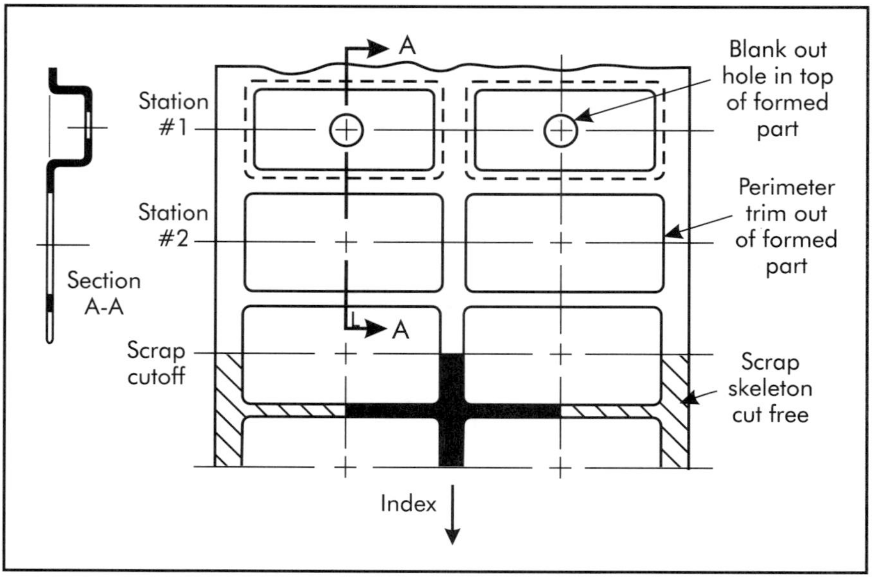


*Figure 11-4. A progressive die with in-line workstations trimming a hole and then the perimeter of a formed part.*

## Trim Press

The die designer should study the workings of each type of off-line trim press and its feeding system to provide necessary services for some of the following conditions:

- Fastening the tool into the proper location on the trim press requires care in laying out die-set mounting holes. When a horizontal trim press is utilized, its large-diameter mounting bolts support the whole weight of the die set and they can not be easily shifted if misaligned.
- Orienting the die-cavity exits to the press opening permit the trimmed parts to travel without impediment out of the die through the existing press opening in the stationary platen.
- Stack parts for easy removal and counting by using trays or custom-designed chutes for each different type of part exiting the die-set area.

- Auxiliary guides and pushers are needed to locate and guide rows of parts directly into the trimming area of the die (as shown in Figures 5-9 and 8-6). Press guides and other mechanisms should be prepared for each thermoformed part to reduce setup time at the press. Many presses require specially designed cams to drive the press feed stroke and provide an accelerating or decelerating motion for smooth movement of the indexing system and the web.
- A knockout device can be used to push cut parts down into the die cavity so that they do not pop out of the die and interfere with incoming indexed shot. These knockouts are attached to a mechanical device built into the trim press, which is triggered to act after the parts are cut free.
- Die clearance should be provided for index guides that may intrude into the die area and for those portions of the formed web that may overlap die buildups. Incoming rows of parts may require clearance built into the die plates to avoid being damaged when the press closes.

The die assembly is the most important tool component in the trim system, but intelligent and timely design of many of these auxiliary mechanical devices determines whether the die changeover can be done in a matter of hours or days.

## Maximum Cutting Force of Press

The maximum rated cutting force of the press is available only at the maximum rpm of the press flywheel. Most press flywheels are directly driven by means of a variable-speed motor. This ability to adjust the rpm of the flywheel (and the press strokes per minute) helps the trim press coordinate its output with the thermoforming machine. Trim-press stroke length can be adjusted to allow varying depths of parts to be indexed into the open die space. The press rpm is reduced as the stroke is lengthened because the platen velocity and its kinetic energy increase, which may cause press or die damage.

## *Tonnage Rating*

The trim-press tonnage output is rated by the press manufacturer at a specific rpm and press stroke. A flywheel stores kinetic energy at the rated rpm and delivers it through its crankshaft stroke at its shut height as a force (rated tonnage). Looking at equation,

$$K_e = CF_r^2 \tag{11-1}$$

where:

$K_e$ = kinetic energy, ft-lb (M-kg or Joule)
$C$ = constant gravitational force (32.2 ft/sec), radius, and weight, which are uniform at any revolution speed
$F_r^2$ = flywheel rpm

If the rpm is reduced below the manufacturer's rating, the output tonnage is proportional to the slower rpm$^2$ when compared to the rated rpm$^2$. For example:

Rated press tonnage = 35 tons (70,000 lbf [311 kN]) at 100 rpm

Find the number of tons (kN) available at 33 rpm.

where:

$$\text{rpm tonnage} = \frac{\text{slower rpm}^2}{\text{rated rpm}^2} \times \text{rated press tonnage}$$

Therefore:

slower rpm = 35
rated rpm = 100
rated press tonnage = 35

$$= \frac{33^2}{100^2} \times 35$$

$$= 3.81 \text{ tons } (7{,}620 \text{ lbf } [34 \text{ kN}])$$

## Cutting Force

The cutting force required for trimming plastic in a punch and die is based on this relationship:

$$F_C = S_M A \tag{11-2}$$

where:

$F_C$ = cutting force, lbf (kN)
$S_M$ = maximum shear stress, psi (kPa)
$A$ = area to be cut, in.$^2$ (cm$^2$)

**Shear stress.** The $S_M$ factor must be deduced empirically because standard plastic physical property tables give tensile stress at break point, but no shear-strength values. An $S_M$ factor for commodity plastics that may be used is 5,000 psi (34,500 kPa), which includes a respectable safety factor. For example:

> Find the cutting force required to trim a single row containing six cavities with a 30-in. (76.2-cm) long cut-off blade and a stripper plate exerting 10 lbf/in. (1,745 N/m) of trim length. The material is 0.06-in. (1.5-mm) thick high-impact polystyrene (HIPS) and each cavity is 4 × 8 in. (101.6 × 203.2 mm).

$$F_t = F_{tp} + F_{sp} = S_M A_{tp} + 10P_t \tag{11-3}$$

where:

$F_t$ = total force to trim shot, lbf (N)
$F_{tp}$ = cutting force to trim cavities and trim scrap, lbf (N)
$F_{sp}$ = force to clamp shot with stripper plate = 10 lbf/in. (1,745 N/m)
$S_M$ = maximum shear stress of plastic = 5,000 psi (34,500 kPa)
$A_{tp}$ = total trim area of cavities and trim scrap, in.$^2$ (cm$^2$)
$P_t$ = total trim perimeter of cavities and trim scrap, in. (m)
$P_c$ = perimeter of a cavity, in. (cm)
$N$ = number of cavities
$L$ = length of trim scrap, in. (cm)
$T$ = thickness, in. (cm)

Therefore:

$$P_c = 2(4 \times 8) = 24 \text{ in. } (61 \text{ cm})$$
$$P_t = NP_c + L$$
$$(6)(24) + 30 = 174 \text{ in. } (442 \text{ cm})$$
$$A_{tp} = TP_t = (0.06)(174) = 10.44 \text{ in.}^2 \ (67.4 \text{ cm}^2)$$

Then:

$$F_t = S_M A_{tp} + 10P_t = 5{,}000(10.44) + 10(174) = 53{,}940 \text{ lbf}$$

(240 kN) = total cutting force

**Minimum rpm.** The die requires 53,940 lbf (240 kN) of cutting force and can be operated in a 35-ton (70,000 lb [311 MPa]) press rated at 100 rpm. The minimum rpm is calculated as follows:

$$R_{min} = \sqrt{\frac{F_l}{F_m} \times R^2{}_{max}} \tag{11-4}$$

where:

$R_{min}$ = minimum rpm
$F_l$ = lesser press force, lbf (kN)
$F_m$ = maximum press force, lbf (kN)
$R_{max}$ = maximum rpm

Therefore:

$$R_{min} = \sqrt{\frac{53{,}940}{70{,}000} \times (100)^2} = 88 \text{ rpm}$$

If 88 rpm is too fast to synchronize production with the thermoformer, two options are available:

1. A trim press running at maximum rpm can be stopped automatically by sensors that are energized when the indexing web coming from the thermoformer becomes taut. The sensors can restart the press when the web becomes slack. However, starting and stopping the press can cause misregistration of the web on each press startup.

2. Stepping the height of the punches so all of the punches do not contact the web simultaneously can reduce the cutting force required to trim (see Figure 11-5). Not all models of trim presses and feeds operate as described earlier and the die designer needs to consult the trim press manual for guidance.

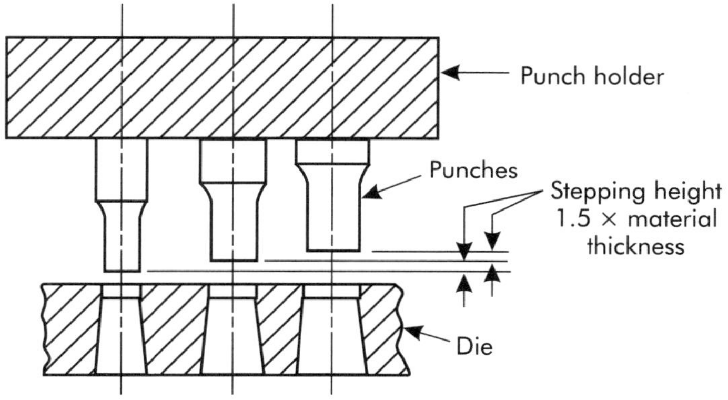


*Figure 11-5. Method of reducing cutting forces by allowing punches to cut consecutively (Nee 1998).*

## Punch-and-die Tooling

### Tool Clearance

The punch-and-die components are the centerpiece of the tooling package. Clean shearing of a thermoformed part requires that the die clearance (the space on one side, between the die and the punch) not be large enough to cause the plastic to bend and break. Conversely, clearance should never be so slight that two tool components are able to rub against each other to cause metal gouging and tool wear.

Other factors loom large on the question of selecting a clearance for the tool such as weight, operating conditions, skills, die-handling methods, and type of plastic to be trimmed.

## *Weight*

The weight of the horizontal trim-press moving platen and the punch holder acts downward with gravity to squeeze all of the moving platen and die-set running clearance to one side of the punch and die. This problem does not exist in a vertical press since gravity does not act on the tool in the same manner to affect running clearance. Many die makers do not have the equipment to test the die in the horizontal position. The die user should be given options to adjust clearance in the tool later on site if needed.

## *Operating Conditions*

A die designer cannot control the operating conditions of a horizontal press platen. However, specifying the largest practical diameter pins, tightest tolerance bushings, and using a four-post die set to offset gravity may help alleviate the action of a worn press.

## *Skills*

The skill and ability of the die shop personnel to build the tool according to specifications are important variables in selecting a die clearance. Punches and dies may be machined exactly to specifications. Yet, if they are not mounted concentric with respect to each other, clearance will not be uniform around the cutting-edge perimeter.

Many die shop personnel developed their skills and knowledge while building metal forming and trimming dies. The vast majority of these metal-fabricating dies weigh less than 300 lb (136 kg) and are set up in vertical presses, which are very rigid and designed with long, precision dovetail-guided slides. Horizontal thermoforming trim-press dies weigh 1,000–2,500 lb (454–1,134 kg) and are set up on massive platens, which are not as rigid as metalworking machines. Moving press platens are often guided by sliding bushings running on shafts and partially supported by platen-to-press slides, which may require skillful adjustment.

Newer models of horizontal trim presses are available with V-ways. These presses reportedly support the die properly and may reduce the gravity effect.

## Die-handling Methods

When metal-stamping die shops first started fabricating horizontal thermoforming trim dies, the size and weight of the dies required a change in die-handling methods and an alternative approach to die building. Selecting die clearance became a balancing effort between too little clearance that can create the conditions for a chipped die, and excessive clearance, which can yield poor trim quality.

## Type of Plastic

Die clearance selection is influenced by the type of plastic resin to be trimmed. The most clearance is specified for the more brittle materials and the least for more rubbery plastics (see Table 11-2).

Table 11-2. Resin comparative die clearance

| Material | Clearance |
|---|---|
| High-impact polystyrene (HIPS) | Most |
| Polyvinyl chloride (PVC) | ↓ |
| Acrylonitrile butadiene styrene (ABS) | to |
| Polyethylene terephthalate glycol (PETG) | ↓ |
| Polypropylene (PP) | ↓ |
| High-density polyethylene (HDPE) | Least |

Options are available to the die user to reduce excessive clearance in the die after it is installed in a horizontal trim press. These alternatives can be utilized after die delivery to improve performance, but they are not a substitute for skillful workmanship, thoughtful design, and high-quality die components.

## Fastening Punches

Many die users prefer to fasten the punches into the tool after the die set is mounted in a horizontal press and gravity has had its effect on the tool. When the moving platen is closed to the cutting position, the punches are inserted through the die opening, which is directly accessible through the back of the press. The punches are loosely held in place by bolts while thin shim stock is used to centralize tool clearance. When the punch is centered, fasteners are tightened to full torque specifications. The die cavities are rigidly bolted and doweled in place to prevent them from shifting during the running of the press. This procedure allows the die and platen weight to be negated so that uniform tool clearance can be preserved.

Die purists may detect some heresy in not mentioning installation of dowel pins to locate the punches. If dowel pins are inserted at the die shop, they will always shift the punch to a preset position, which may or may not be correct for a horizontal press setup. Since bolts are the only fasteners used for attaching the punch, a sufficient number must be employed to mount and maintain the punch to its correct position. Before the tool is taken out of the trim press and placed in storage, the punches can be removed to protect them from damage. Alternatively, if the die set (with punches still mounted) is removed from the press, it should be blocked from closing to its cutting position when stored.

## Hardness

Some die makers recommend having the die cavities heat treated to less than full hardness and the punch to maximum hardness. This difference in heat treatment allows the tool user to peen (displace) metal inward toward the die cavity to remove any excess wear, which occurs during thermoforming production. When the press is closed manually, the hardened punch shears through the peened section of the softer die cavity, providing a fresh cutting face. This avoids shutting down the thermoforming line for die maintenance during production.

It is important to understand that the words "softer" or "less-than-full hardness" are relative, since a die must be hard enough to maintain a cutting edge during its reasonable production life. Tool steel tempered to its recommended maximum hardness can be somewhat brittle. Tempering the steel at a higher temperature tends to make the steel tougher (less brittle) but also less hard. A study of the tempering chart for tool steel may indicate the point that is a compromise among the best qualities of the steel. If both punch and die are fully hardened, care must be exercised that they always have sufficient clearance when being closed to avoid chipping the tool.

A die is considered dull when its cutting edge shows evidence of a radius developing during production trimming. Grinding the hardened tool face (never on its perimeter edge) to eliminate a die-edge radius will resharpen the die or punch.

## Die Sets

The die set is a complete tool module that contains the punch and die and other necessary components for trimming thermoformed parts from the web (see Figure 11-6). Commercial off-the-shelf die sets are rarely adaptable for thermoform trimming. Custom-built die sets are either purchased or fabricated by the tool shop. Because of their weight and since they often operate in the horizontal position, automatic lubrication of guide-pin bushings is essential. Contamination of the bushings by dirt or dust can cause them to seize or gall. They should never leave their guide pins when the moving platen is fully retracted.

## Guide Pins and Bushings

Different types of guide bushings are commercially available. The best guide pins and bushings create the least running friction and have minimum clearance between the guide pin and bushing. Guide pins and bushings come in two basic styles: removable and those that are press-fitted into the die set (see Figures 11-7

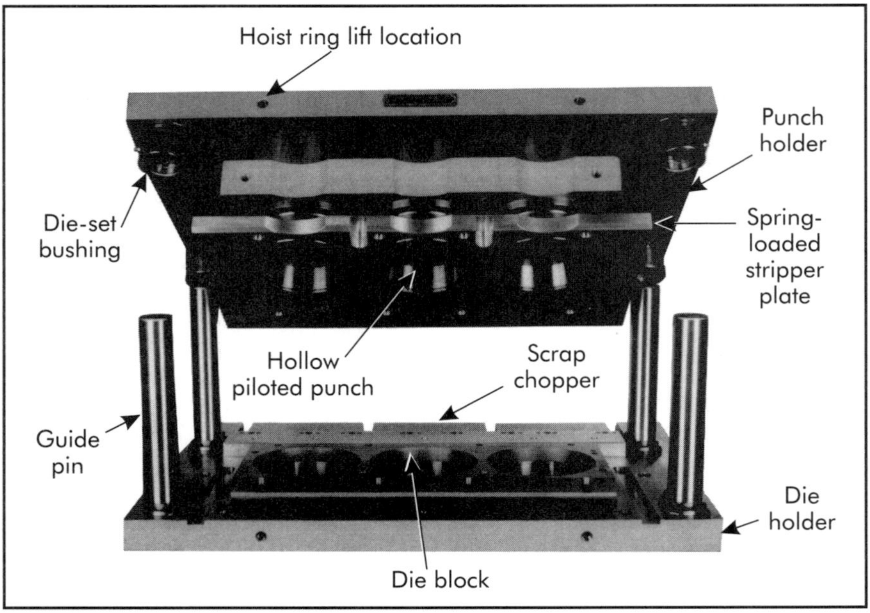


*Figure 11-6. Typical punch-and-die assembly. (Courtesy E.D. Segen & Co., LLC)*

*Figure 11-7. Typical die-set bushings. (Courtesy Producto Machine Company)*

and 11-8). All pins are hardened and ground and can be purchased in standard diameters of any length. Removable guide pins are seated in conforming tapered bushings or precision straight holes to maintain true alignment. They are easily disassembled to facilitate tool fabrication and sharpening.

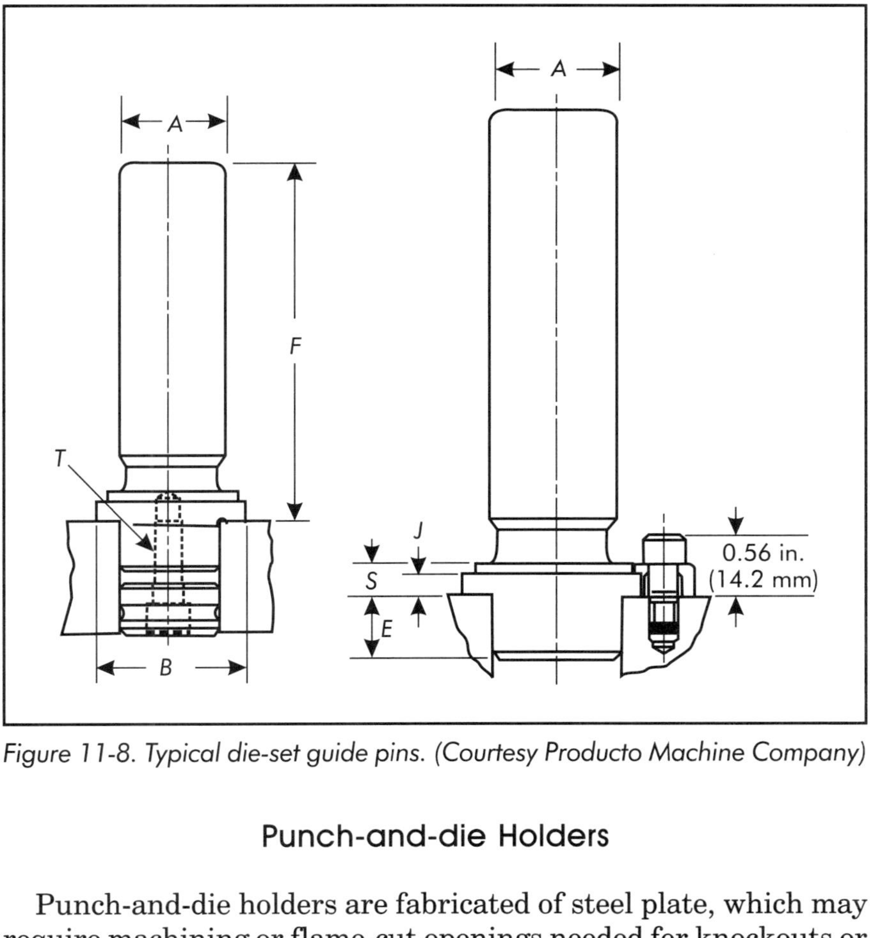


*Figure 11-8. Typical die-set guide pins. (Courtesy Producto Machine Company)*

## Punch-and-die Holders

Punch-and-die holders are fabricated of steel plate, which may require machining or flame-cut openings needed for knockouts or to act as an exit for the trimmed thermoformed parts. These plates are stress relieved by special heat treatment. Their parallel faces are ground flat, and the four plate edges machined square and to size. It is important for additional precision machining of these components that the plates are relieved of internal stresses that can cause warping, bending, or other distortions. Buildup plates used in the fabrication of the die are similarly processed.

## Stripper Plate

A stripper plate is an essential tool component, which is spring-loaded and located on the punch holder and set 0.06–0.13 in. (1.5–3.3 mm) in advance of the punch. Initially on press closure it clamps the web to the die face, preventing the plastic sheet from shifting as the punch enters the die cavity. After the parts are cut, it forces the web off the punch, freeing the shot for the next index. The compression force of stripper springs should be added to the required trimming force. The combined amount should be within the rated press capacity.

## Pilots

Pilots are components of the punch and can be either male or female configurations shaped to the contour of the formed plastic cavity (see Figures 11-9 and 11-10). They may be separately attached or machined into the body of the punch so that each cavity is individually located by its own pilot. The trim-press index system locates a row of parts into the die area where the pilots pick them up to complete precise registration.

Auxiliary cone locators are sometimes used to correct the index of a row of parts prior to the pilot engaging the sheet (see Figure 11-11). These spring-loaded locators are designed to nest in pre-thermoformed conical portions of the web.

## Knockouts

Knockouts are used to push trimmed parts out of the web down through the die, providing clearance to allow indexing the next row of parts into the die area. Female punches contain an internal cavity that acts as a pilot for the formed part extending inward toward the punch. The cut component remains up above

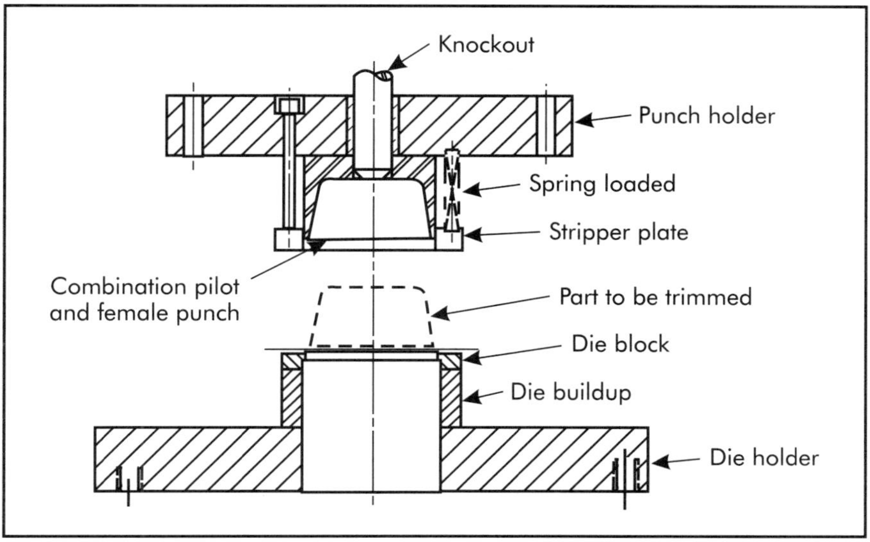


*Figure 11-9. Combination pilot within a female punch.*

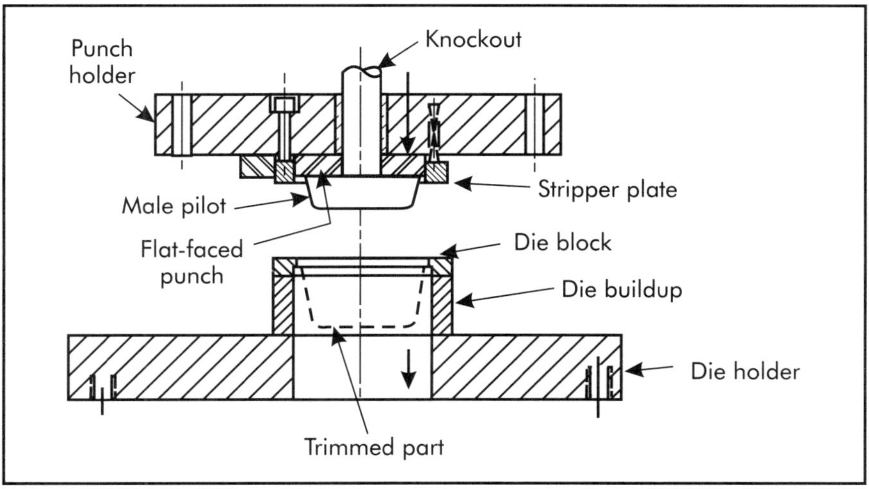


*Figure 11-10. Male pilot mounted on a flat-faced punch.*

the die face until the knockout action forces it below the sheet line (see Figure 11-9). Flat-faced punches can be used to trim components that have their formed depth facing into the die cavity. The knockout then pushes cut parts past the die lands (see Figure

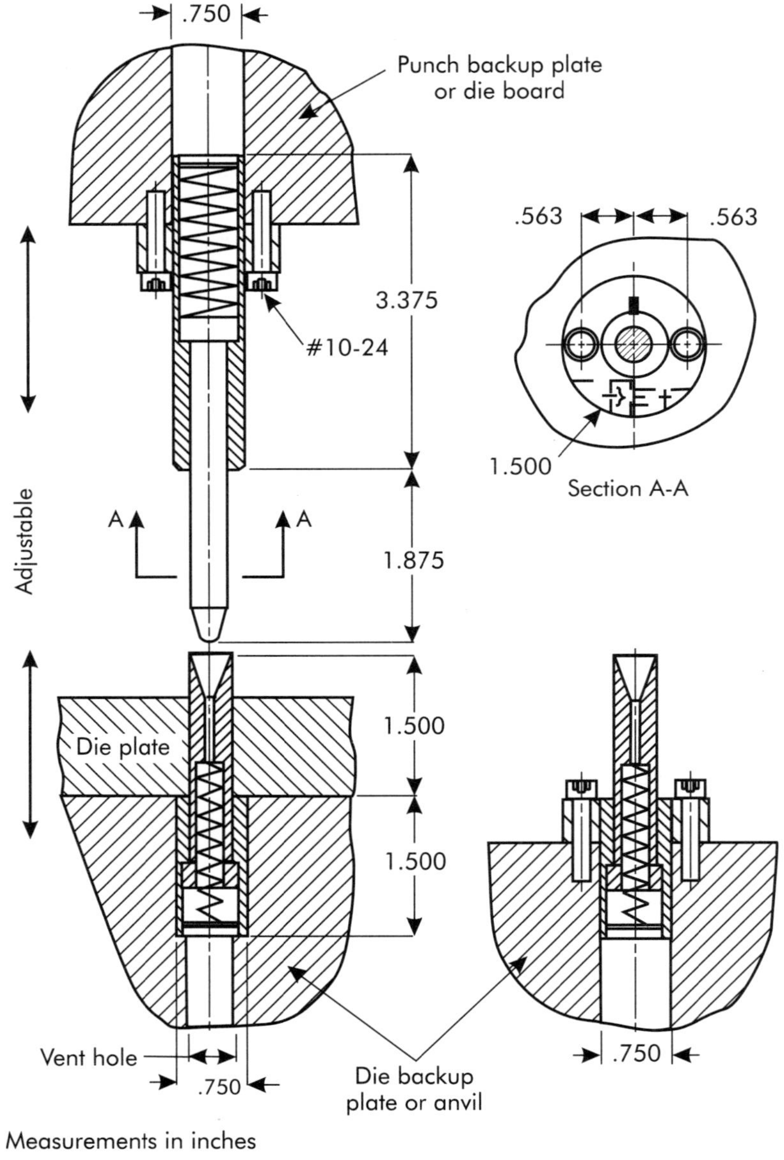


*Figure 11-11. Auxiliary cone locators to assist registration of a row of parts relative to the die. (Courtesy E.D. Segen & Co., LLC)*

11-12). The die lands are the precision-machined portion of the die that acts as the trimming area. Beyond the land area, clearance allows cut parts to move freely through the die. Normally, the die land is only 0.19–0.32-in. (4.8–8.1-mm) long and the punch usually penetrates the land area during the cutting process by approximately 0.03 in. (0.8 mm). This is to avoid undue wear on working portions of the tools and to allow room for resharpening (see Figure 11-12). Finished cut parts left in the land can be jarred out of the die land area by press vibration. The knockout serves to avoid these misfeeds.

Die backup and buildup plates are needed in a die for spacing the web away from punch-and-die holder plates (see Figure 11-12). Clearance for the depth of incoming thermoformed parts is provided by the buildup plates to prevent crushing parts when the press is closed. Multicavity punches and dies are often assembled on individual backup plates outside the die set. Completed subassemblies are mounted and later aligned in the die set. This assembly is used to allow relatively low-cost, thicker, mild-steel

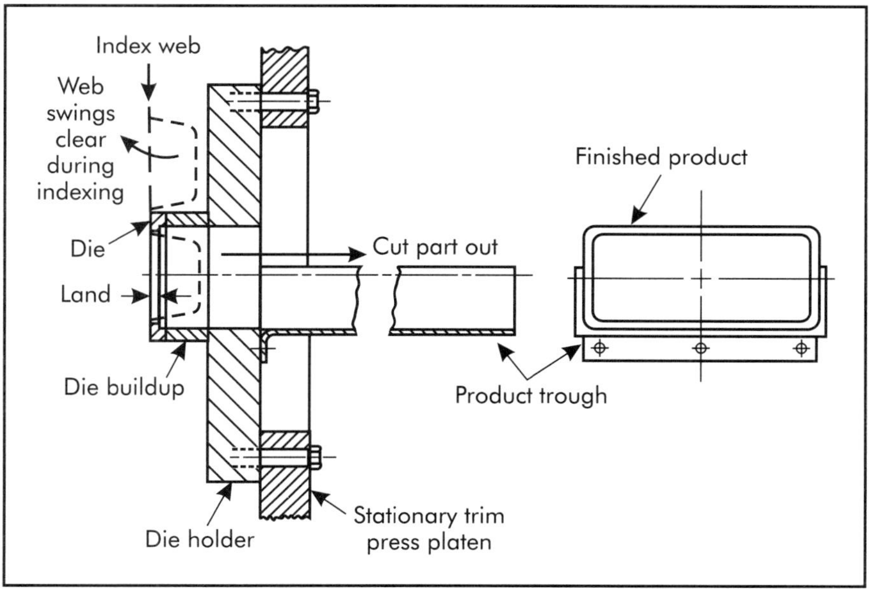


*Figure 11-12. Product trough organizes parts into stacks for counting and packing.*

plates to back up and support the thinner, expensive hardened tool-steel, punch-and-die components and to clear the formed part height.

## Exit Chutes or Product Troughs

Exit chutes or product troughs help organize trimmed-part output into stacks for counting and packing (see Figure 11-12). Efficient production may be disrupted if the chutes are not planned for the specific trimmed parts, leaving the user to contrive a makeshift setup.

Troughs should be carefully aligned to the die-set exit cavities to provide a smooth part transition from die set to trough. Very high-production items such as cups or lids may be counted, sleeved in polyethylene ("poly") bags, and packed by special automatic equipment. Shorter production runs of parts are manually counted by measuring a stacked height of trimmed components and stripping them directly off the trough of a horizontal trim press into a polybag. The length of the chute should be three-to-four times the length of a stack of packed products. Horizontal press-trimmed parts advance out of the die exit with each stroke of the press and will stack properly when a sufficient number of components are left on the trough to lightly compact the incoming product stream.

Troughs are best fabricated of stainless-steel sheet metal rather than mild steel, because stainless steel does not need painting to prevent rust. Paint or rust increase frictional resistance to part movement and may contaminate food or medical components that require cleanliness and sanitation.

## Cutoff Blades

Scrap cutoff blades, which trim the indexed web with every stroke of the press, are necessary for a horizontal trim-press die (see Figure 11-13).

The decision to use a simple straight cutoff blade or to trim the scrap into two or more pieces depends on several factors that

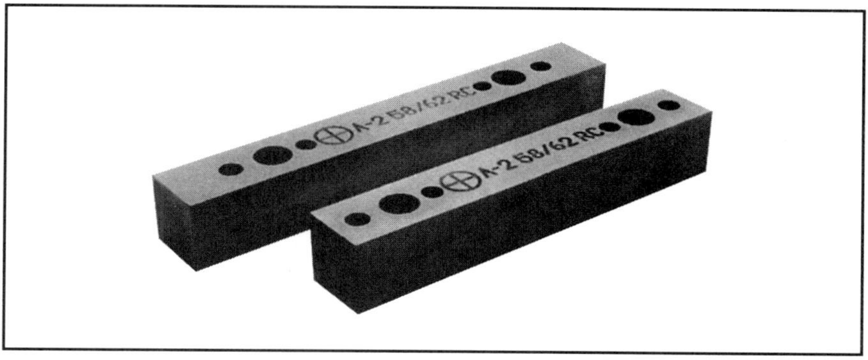

*Figure 11-13. Commercially available scrap cutoff blades hardened and ground to size can be combined to increase length. (Courtesy E.D. Segen & Co., LLC)*

come into play when granulating in-line, trimmed scrap (see Figure 11-3).

- The granulator may not have sufficient shredding capacity to accept a straight cutoff scrap web, yet may be capable of processing the same skeleton when cut into segments by the die. Lack of grinding capacity can be due to either a high-cyclic rate of output from the trim press or inadequate granulator shredding horsepower. Granulators may have small-inlet throat openings that limit the size of the skeleton, which can cause jamming or bridging of incoming full-width web scrap at the inlet throat.
- Die set available length in the index direction can be a limiting factor in the design of the web cutoff when it is used to trim the scrap skeleton into several pieces. Trimming scrap into segments requires the web to advance one additional index beyond the trim cavity to allow sufficient space to mount the scrap cutoff blades.
- Additional die cost is incurred to provide multiple-segment cut-up of a scrap skeleton.

The die user should have the production experience to guide the tool designer in providing the scrap-cutoff method most suitable for the user's scrap-handling equipment.

## Safety

Die-handling safety devices should be made available with each die with instructions for their use when moving dies from storage to the trim press. Off-line horizontal trim-press configuration requires that a die set be lifted up into the press by chains attached to eye bolts, or hoist rings fastened to the edges of the die-set plates. These lift devices should be fabricated of high-quality forged steel and their rated capacity should be drastically reduced if the chain is set at an angle to the vertical axis. For example: if a chain when loaded rests at 45° (the maximum allowable angle) to an eyebolt, its lifting capacity is reduced to 25% of its rating. Many types of hoist rings allow 360° swivel and 180° pivoting, which helps prevent eye-bolt breakage from side loads or twisting. Full rated load is permitted if all the manufacturer's specifications are observed in the use of these safety devices.

When removed from the press, the die set should have a secure system to fasten the movable punch holder to the die holder plate to prevent accidental opening or closing while being handled.

## Thermoforming Molds and Trim-in-place Punch and Dies

Thermoforming molds and trim-in-place punch and dies are a combination tool, as shown in Figures 11-14 and 11-15, which require a special thermoforming machine. These tools eliminate the need for indexing a thermoformed part from the mold to a separate punch-and-die trim press. Perimeter-trimming registration relative to the formed part is the most accurate obtainable in any existing system. These tools and equipment are very popular throughout the world and used for many special projects in North America.

Thermoforming with trim-in-place punch and dies uses either vacuum or pressure forming and plugs when needed. Finished thermoformed parts are mechanically ejected directly into individual stacking chutes. Thermoforming is mainly accomplished with female molds mounted within the punch, although shallow male cavities, such as lids, can be directly formed using this process.

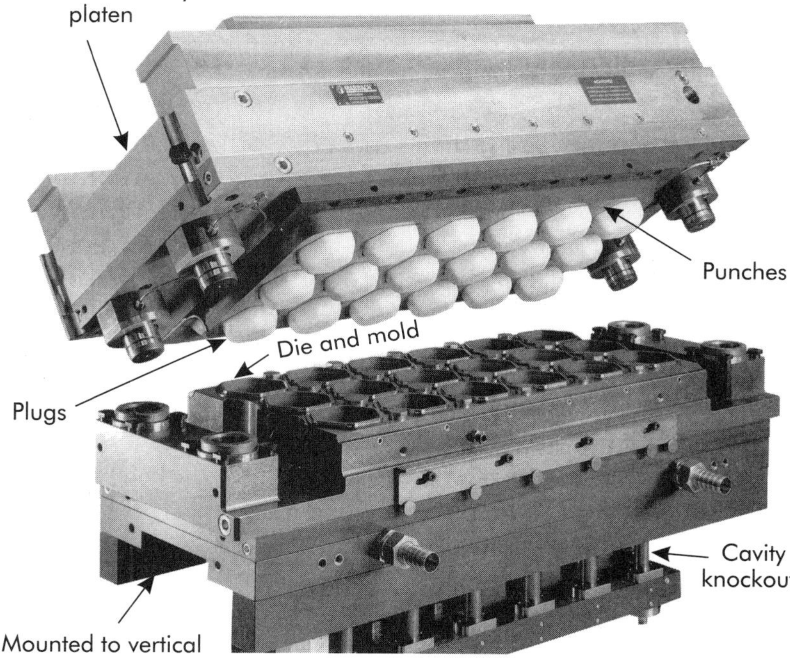


*Figure 11-14. Tilting trim-in-place tool. (Courtesy Marbach Werkzeugbau GmbH)*

In vertical presses, the punch platen is stationary and the die platen travels vertically. The weight of the tooling and press platen does not affect the die bushing and pin clearance, as it does in a horizontal trim press. Removal of finished trim parts requires the bottom platen to rotate and tilt at an angle to align itself with a stacking cage. Tilting the lower tool causes both die-set bushings and lower-press-platen bushings to leave their guide pins. Disengagement of pins and bushings may cause excessive wear. Dust and dirt accelerate this process. Excessive wear in the die-and-platen guidance system may increase allowable tool clearance and eventually affect the trimming ability of the tool. The trim-in-place

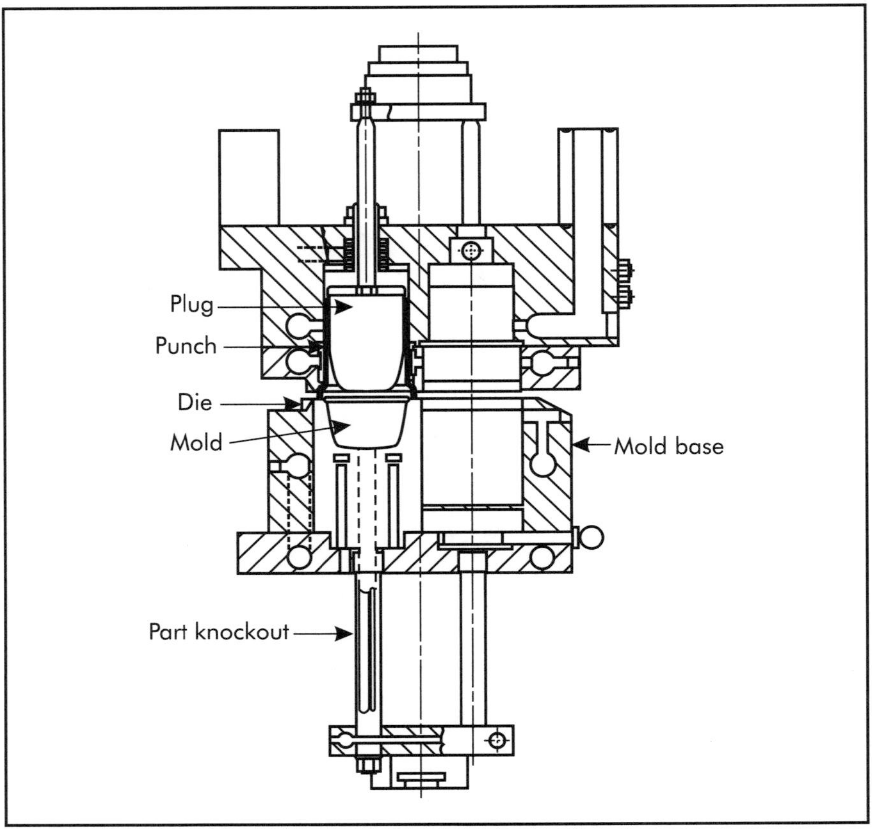


*Figure 11-15. Thermoforming trim-in-mold tool (Mitchell 1996).*

die designer may specify more guide pins and bushings than are commonly used on dies that never leave their guide pins.

When thermoforming and trimming take place in the same tool, heat from the web can affect trim-tool clearance. A temperature differential control system is available that maintains the moving and stationary die-set platens within 5° F (3° C) of each other to avoid damage to the die cavities due to heat expansion. The temperature-control system can be programmed to halt the machine if the temperature differential exceeds specifications.

The economics of a trim-in-place system require study of the cost of providing a punch-and-die to accompany each thermoforming cavity in the mold. The cost per cavity tends to limit the number of cavity rows specified for many projects. The cost per die row may become an important economic factor when comparing the cost of a trim-in-place die to a horizontal, off-line, trim-press die (which needs only one or two rows of punches and dies to trim a multi-row mold). Each of these methods of production has approximately the same mold cost for a similar number of cavities. The cost for the additional part-stacking chutes, knockout devices, and differential die-temperature system further increases the price of a trim-in-place tool. However, the tooling cost per unit of a trim-in-place system becomes economically viable when large numbers of plastic components of the type best suited for trim-in-place tooling are required.

## Layout of an Off-line Punch and Die

Mold layout, the plastic resin specified, input gained through customer consultation, and trim-press data all influence punch-and-die design.

The following steps are used to gather specific design information.

1. Determine the number of rows of trim tools required.

$$N = \frac{S}{C} \tag{11-5}$$

where:

$N$ = number of rows of trim tools required
$S$ = strokes/min required to trim web = thermoformer's output (shots/min) × number of rows in one shot
$C$ = maximum trim-press strokes/min output (per manufacturing specification)

2. Access the resin shrinkage coefficient from Table 8-1. Use full shrinkage, since the web cools to ambient temperature by the time it reaches the die.
3. With the trim press operating at rated speed, calculate the estimated tonnage and compare it to the cutting force required to trim (see Equations 11-2 and 11-3). If press tonnage

is insufficient, consider stepping punch heights by allowing trimming to take place consecutively (Figure 11-5). Alternatively, the punch cutting face can be ground at a shear angle so that the punch does not trim the whole die perimeter at once but instead cuts incrementally.

4. Decide whether the scrap web should be cut into multiple segments or cut off in one piece, based on scrap-skeleton granulating capacity.
5. Decide on an appropriate punch-and-die tool clearance.
6. Decide whether the die will be peened while in service. Select suitable tool steel and heat treatment for the punch and dies.
7. Select the type of guide bushings and pins to be specified for the die set.
8. Resolve whether the customer or die maker will supply auxiliary devices, such as output troughs, web guides, and indexing pushers.
9. Obtain a trim-press-platen layout drawing and specifications to locate the die set on the press platen and check the following items:
   - die-set mounting holes,
   - guide-pin clearance locations through the moving press platens,
   - knockout location on moving-press platen opening,
   - stationary platen trimmed part exit opening, and
   - standard adjustable press stroke and shut-height adjustment specifications to determine pin guide length.
10. Determine if the thermoformed section will be facing the punch (female punch) or protruding into the die (flat-face punch).
11. Determine if the die will be designed to trim an interior segment and then index again within the die set to blank out the finished part (see Figure 11-4).
12. Determine if the end user is to install the punches after the die set is mounted in a horizontal trim press or if they will be dowel-pinned rigidly in place on the punch plate by the die maker.
13. Determine if the end user requires cone locators to assist in locating each indexed trim row.

14. Determine the requirements for special types of lift eye bolts, hoist rings, or other methods of preventing opening and closing of the die set while it is being mounted in the press.

## Design Parameters

Table 11-3 outlines the die design specifications for the 18-cavity mold shown in Figure 11-16.

The design of the mold outlined in Table 11-3 is such that the die can be built with either three or six cavities in each transverse row. The cost of a three-cavity die might be 50% less than the price of a six-cavity die. In this case, economics call for the least number of cavities.

**Table 11-3. Die design specifications for an 18-cavity mold**

| Specification | Parameter |
|---|---|
| Plastic web—high-impact polystyrene | 0.04-in. (1.0-mm) thick<br>30.75-in. (78.1-cm) wide |
| Mold area | 29-in. (73.7-cm) wide × 28.21-in. (71.7-cm) long |
| Perimeter trim size—no interior segments pre-trimmed | 4 × 8.63 in. (10.2 × 21.9 cm)<br>0.25-in. (6.4-mm) corner radii |
| Flat punch and male pilot | See Figure 11-10 |
| Maximum off-line trim-press speed with a 4.5-in. (11.4-cm) stroke is 100 strokes/min at 60,000 lbf (267 kN) | See Table 11-5 |
| Thermoforming output | 16 shots/min |
| Number of cavities | 18 = 3 across × 6 rows deep per shot |
| Depth of part | 1.50 in. (38.1 mm) |
| Scrap cutoff blade—straight blade includes excess for web misalignment | 32.25-in. (81.9-cm) long |
| Stripper plate clamp force | 10 lbf/in. (1,745 N/m) of trim perimeter |

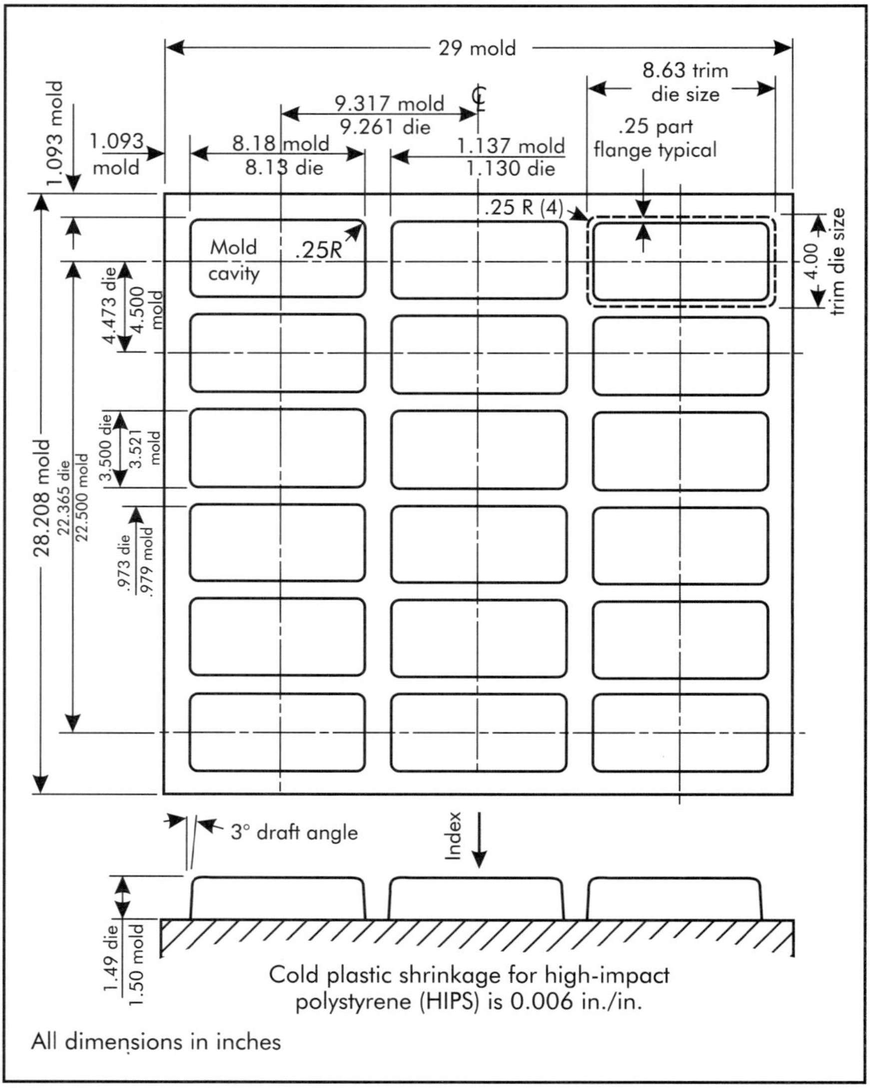


*Figure 11-16. An 18-cavity mold layout.*

First, determine if the complete output of the thermoformer is capable of being processed in a die containing three cavities in a row. Then, Equation 11-5 can be used to determine the number of rows of trim tools required. Assume:

$S = 16 \times 6 = 96$
$C = 100$

$$N = \frac{96}{100} = 1 \text{ row of trim tools}$$

So, one row of three cavities will suffice.

Thermoforming certain resins or shapes of parts can be adversely affected by the extruder direction relative to the geometry of the mold cavity. As this particular mold is not sensitive to extrusion direction, three cavities in a row would be suitable for this die layout.

## *Trim Force*

The trim-press cutting force must be sufficient to trim three cavities and compress the stripper-plate springs. Calculate the perimeter of the three cavities and length of the web cutoff blade. Using Equation 11-3, assume:

$P_c = 2(4 + 8.63) = 25.26$ in.
$P_t = 3(25.26) + 32.25 = 108$ in.
$A_{tp} = TP_t = 0.04(108) = 4.32$ in.$^2$

Therefore:

$$F_t = S_M A_{tp} + 10P_t$$
$$= 5{,}000(4.32) + 10(108) = 22{,}680 \text{ lbf (101 kN)}$$

Rated press force at maximum rpm of 100 rpm = 60,000 lbf (267 kN)

The available press force is more than double the total trim force needed to cut three parts in a row.

## *Die Clearance*

Polystyrene plastic is somewhat brittle so that it will tolerate a more generous clearance than a more rubbery plastic. One method for calculating the die clearance is to enlarge the gap between the punch and die to compensate for an increase in sheet thickness. A tool clearance of 3% of sheet thickness is a good starting point. From Table 11-2, the high-impact polystyrene (HIPS) sheet thickness is 0.04 in. (1.0 mm) × 3% = 0.0012-in. (0.030-mm) tool clearance (per wall).

A more rubbery plastic (high-density polyethylene) may require 1.5–2% tool clearance or 0.0006–0.0008-in. (0.015–0.020-mm) tool clearance.

To avoid excess tool wear, a minimum of 0.0003–0.0005 in. (0.008–0.013 mm) is a practical lower limit for tool clearance.

## *Guide Pins*

Most die sets use the four-guide-pin design (Figure 11-8) but some older presses may accommodate only two-guide-pin tooling. The pins must be of sufficient length so they are fully engaged by the guide-pin bushings at the maximum working opening of the trim press. Guide pins are hardened, ground, and mirror-finished to 5–8-μin. (0.1–0.2 μm) for maximum life and to reduce frictional resistance. Guide pins can be press-fitted into the die holder for maximum accuracy and squareness. The removable-style pin is specified for ease of die fabrication, sharpening, and for rapid replacement of worn pins. Special guide pins, which are used with pre-loaded ball bearing retainers and bushings, are available as completely packaged assemblies.

## *Guide Bushings*

Guide bushings (Figure 11-7) are available in press-fitted or removable styles and in materials such as hardened steel, tough aluminum bronze, or bronze-plated steel. Nonferrous materials are specified to minimize damage that could occur if the lubrication system fails, as steel-to-steel components may seize and gall.

Pre-loaded, ball-bearing bushings have a retainer cage in which the steel balls are arranged in a helical pattern so that each ball has a separate linear path to travel on the guide pin (see Figure 11-17). The balls are pre-loaded between an outer steel bushing and the guide pin. This arrangement tends to minimize the clear-

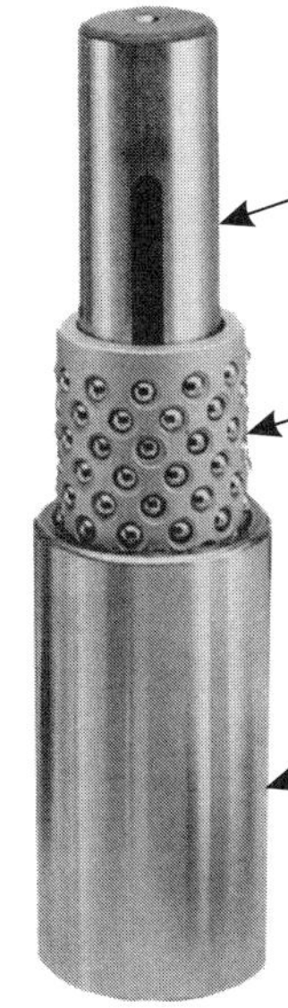


*Figure 11-17. Packaged, pre-loaded, ball-bearing retainer, bushing, and pin. (Courtesy Producto Machine Company)*

ance required for linear travel of the punch plate and provides the most precise lineup for the punch and die.

Selection of a suitable style pin and bushing depends on the allowable clearance between the punch and die, and experience the end user has had with these components. The diameter of the pins and bushing for a four-pin die set range from 1.25–2.00 in. (31.8–50.8 mm), depending on the thickness of the die holder.

## *Plate Thickness*

The plate thickness of the punch-and-die holder plates in the die set is based on retaining sufficient cross-sectional area to keep them flat and parallel to each other. This becomes the foundation on which to build the tool. The distance between the pins and the size of the cutouts for parts clearance that exit from the die require specification of the proper plate thickness. The die set must support itself as a monolithic structure without distortion, and die buildup plates add to the stiffness of the tool. A good starting point for the thickness of die-set plates is 2.50–3.00 in. (63.5–76.2 mm). This assumes that the plates do not contain excessive cut-out areas and have approximately 40 in. (101.6 cm) between the guide pins.

## *Tool Steel Selection*

Tool steel selection and heat treatment is complicated because no one steel can possess all the qualities to make it an easy choice (see Table 11-4). Water (W) or oil (O) -hardening tool steels are readily available and relatively inexpensive. They are easily machined, but tend to deform when quenched rapidly in fluids after undergoing high-heat-treatment temperatures. Air-hardening (A) steels are heated to red heat and quenched slowly in air, which results in minimal distortion. High-carbon, high-chromium (D) steels are difficult to machine but have the greatest toughness and wear resistance of all the tool steels. The remaining tool steels in Table 11-4 are rarely used in thermoforming trim dies.

To prevent stress-concentration cracking, punch-and-die components that are to be hardened should not have sharp corners. Whenever possible, the designer should try to maintain a uniform cross-sectional area to reduce heat-treatment distortion. Die parts are first machined undersize if they are to be ground later to finished dimensions after hardening. Small tools that have a punch-and-die clearance of 0.0008 in. (0.020 mm) or greater are often machined to finished size when fabricated from air-hardened tool steel. Then, they are heat-treated without further internal grinding. These hardened parts are ground parallel on their faces to sharpen their cutting edges and provide a flat mounting surface.

The majority of hardened die components are fabricated from either A-2 or D-2 tool steel, which have some of the following characteristics.

**Air-hardening die steels.** Type A2 is the most often specified type of air-hardening tool steel. It exhibits minimum movement in hardening and has higher wear resistance than oil-hardening die steels, with equal toughness. Steels A4, A5, and A6 can be

Table 11-4. Comparison of the basic characteristics of steels used for press tools

| AISI* Steel Number | Non-deforming Properties | Safety in Hardening | Toughness | Resistance to Softening Effect of Heat | Wear Resistance | Machin-ability |
|---|---|---|---|---|---|---|
| W1 | Poor | Fair | Good | Poor | Fair | Best |
| W2 | Poor | Fair | Good | Poor | Fair | Best |
| O1 | Good | Good | Fair | Poor | Fair | Good |
| O2 | Good | Good | Fair | Poor | Fair | Good |
| O7 | Good | Good | Fair | Poor | Fair | Good |
| A2 | Best | Best | Fair | Fair | Good | Fair |
| A4 | Best | Best | Fair | Poor | Fair | Fair |
| A5 | Best | Best | Fair | Poor | Fair | Fair |
| A6 | Best | Best | Fair | Poor | Fair | Fair |
| D2 | Best | Best | Fair | Fair | Good | Poor |
| D3 | Good | Good | Poor | Fair | Best | Poor |
| D4 | Best | Best | Poor | Fair | Best | Poor |
| D6 | Good | Good | Poor | Fair | Best | Poor |
| S1 | Fair | Good | Good | Fair | Fair | Fair |
| S2 | Poor | Fair | Best | Fair | Fair | Fair |
| S4 | Poor | Fair | Best | Fair | Fair | Fair |
| S5 | Fair | Good | Best | Fair | Fair | Fair |
| H11 | Best | Best | Best | Good | Fair | Fair |
| H12 | Best | Best | Best | Good | Fair | Fair |
| H13 | Best | Best | Best | Good | Fair | Fair |
| H21 | Good | Good | Good | Good | Fair | Fair |
| H26 | Good | Good | Good | Best | Good | Fair |
| T1 | Good | Good | Fair | Best | Good | Fair |
| T15 | Good | Fair | Poor | Best | Best | Poor |
| M2 | Good | Fair | Fair | Best | Good | Fair |
| M3 | Good | Fair | Fair | Best | Good | Fair |
| M4 | Good | Fair | Fair | Best | Best | Poor |
| L2 | Fair | Fair | Good | Poor | Fair | Fair |
| L3 | Fair | Poor | Fair | Poor | Fair | Good |
| L6 | Good | Good | Good | Poor | Fair | Fair |
| F2 | Poor | Poor | Poor | Fair | Best | Fair |

* American Iron and Steel Institute

hardened from lower temperatures. They have lower wear resistance but better distortional properties.

**High-carbon, high-chromium die steels.** Type D2 is often used for thermoforming dies. It finds wide application for long-run dies. It is deep hardened, fairly tough, and has good resistance to wear, but is more difficult to machine than A2. Steels D3, D4, and D6, containing additional carbon, have very high wear resistance and lower toughness. Steels D2 and D4 can be air-hardened.

## *Sheet Shrinkage*

Shrinkage of the plastic, from the size of the mold down to the finished part, should be analyzed as to how it affects die dimensioning. Heated high-impact polystyrene (HIPS) sheet shrinks 0.006 in./in. (0.006 mm/mm) uniformly in the transverse and index directions. High-density polyethylene (HDPE) and some other extruded resins may have a differential shrinkage relative to the transverse and index direction of extrusion.

The following dimensions affect shrinkage and die design:

- Centerline to centerline dimensions of the cavities are affected across the web (transverse). The index direction is affected only when two or more rows are blanked at one time.
- Pilots that are used to position the formed part to the die must conform precisely to the shrunken component to be effective.
- A part whose formed section is not centered to its perimeter trim requires care when designing its pilot location.
- Cone locators are used to pre-register shot rows on the outer edges of the web. An error in shrinkage of 0.002 in./in. (0.002 mm/mm) on a 30-in. (76.2-cm) wide web would be 0.06-in. (1.5 mm) and may cause misregistration of the row.

## *Stripper Plates*

Stripper plates, mounted on the punch plate and guided by shoulder bolts, are spring-loaded to their fully extended position. Stripper steel-plate stock that has been stress-relieved and ground on all faces is used to fabricate this component. The plate thickness can vary from 0.50–1.00 in. (12.7–25.4 mm), depending on how much material remains after machining the cavity to clear the punches. The material is a low-carbon steel, has no alloying elements, and tends to machine with minimal finished distortion.

Choosing die-set design dimensions requires studying the press specifications of the machine manufacturer, as shown in Table 11-5, and platen drawings (see Figures 11-18 through 11-20). Examining the critical dimensions of these trim-press platen layouts will help rough out the overall minimum size of the die set.

- The four guide pins should have a center-to-center distance of 9 × 42 in. (22.9–106.7 cm) with a maximum pin diameter of 2.25 in. (57.2 mm).
- Alternate rows of mounting holes for the punch plate of 10 and 13 in. (2.5 and 33.0 cm) allow direct fastening of the punch

Table 11-5. Brown Machine T-300 trim-press specifications (Courtesy Brown Machine, LLC)

| | Moving Platen Openings | | |
|---|---|---|---|
| Stroke | B—Minimum Shut, in. (mm) | A—Adjustable Length, in. (mm) | Maximum rpm |
| 4½ | 7½ (190.5) | 16 (406.4) | 100 |
| 7½ | 6 (152.4) | 16 (406.4) | 90 |
| 10½ | 6 (152.4) | 14½ (368.3) | 80 |
| 13½ | 6 (152.4) | 13 (330.2) | 60 |
| 16½ | 6 (152.4) | 11½ (292.1) | 60 |

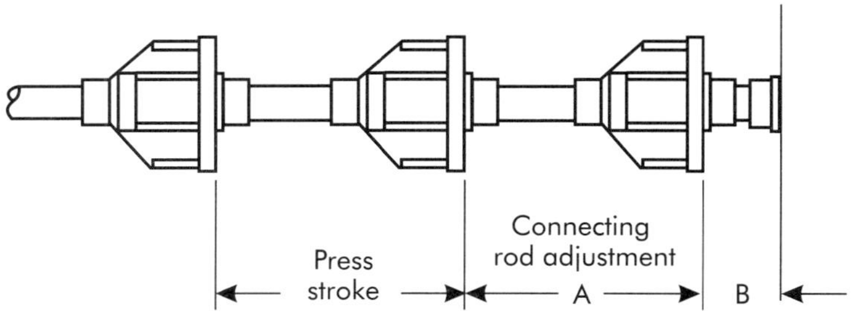

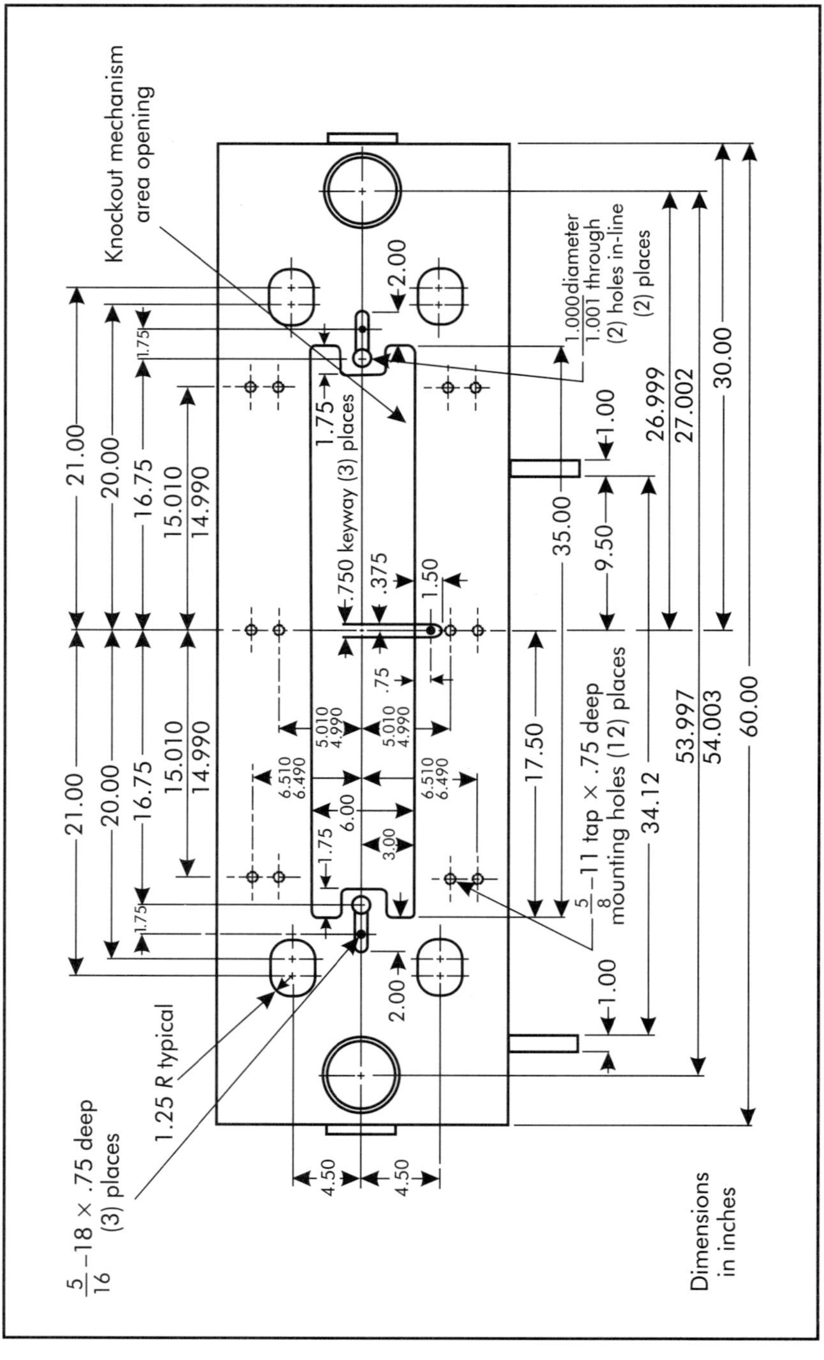


*Figure 11-18. Moving platen layout for Brown trim press. (Courtesy Brown Machine, LLC)*

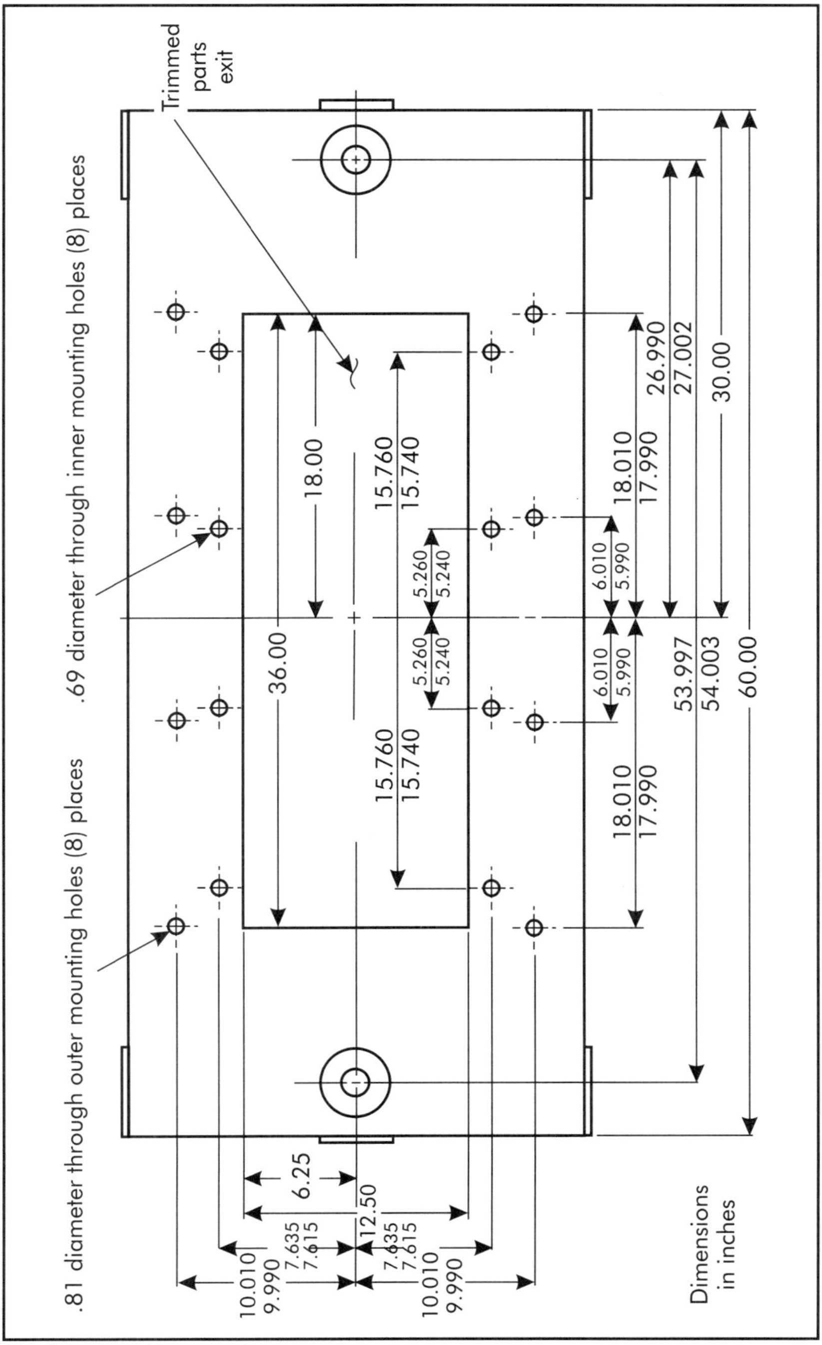


*Figure 11-19. Stationary platen layout for Brown trim press. (Courtesy Brown Machine, LLC)*

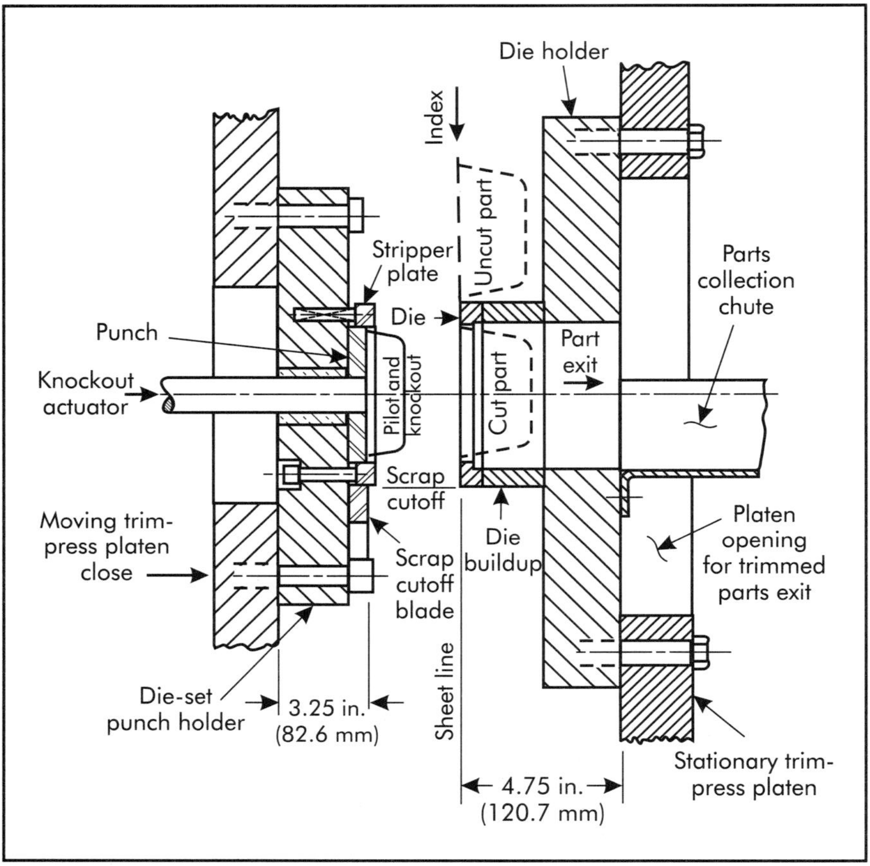


*Figure 11-20. Die shown mounted in a horizontal Brown T-300 trim press. (Courtesy Brown Machine, LLC)*

plate to the moving platen without removing any of the overhanging tooling. The size and weight of the punch plate can be reduced when smaller punches are needed by using the minimum mounting-hole centers.

- The three fixed keys help operators to locate the die set in the correct position on the centerlines of the platens.
- Once the moving platen dimensions are settled, the stationary-platen-mounting dimensions will fall into place.

## Shut Height

The minimum shut height of the press, which is based on the press stroke and maximum rpm, guides the designer as to choice of thickness for the buildup of plates to meet the minimum and maximum specifications (see Table 11-5).

## Sheet Shrinkage

Salient shrinkage dimensions of the mold layout, as shown in Figure 11-16, are recalculated and marked "die" to indicate the cold-shrink size of the plastic to be presented to the trim tool. Die dimensional information should be re-examined for accuracy to provide correct registration of the shot relative to the die. It should be noted that the distance between mold cavities in the transverse and index directions is different. Abutting transverse die-cavity walls will reinforce each other to provide sufficient cross-sectional stiffness to resist deflection by the trim force, and they still can be removed as individual units.

The die user often specifies that a scant minimum of plastic wall be left in the web as trim scrap. A wider die wall may become necessary to provide a strong die and the difference should be resolved by negotiation with the customer. The index directional sidewalls on a single-row die generally have sufficient space available to provide strong die walls.

Figure 11-20 shows platen layouts in the side view of the trim press to help complete the design of a four-post die set. The moving platen has a center opening to allow installing knockouts and 0.75 in. (19.1 mm) keyways at three centerline locations to center the die set. The stationary platen contains a large exit opening that can accommodate parts up to 12.50-in. (31.8-cm) long.

Table 11-5 shows the press stroke and connector rod adjustment for a moving platen. The minimum shut height of the die for a 4.50-in. (114.3-mm) stroke is 7.50 in. (190.5 mm), which sets the minimum dimension for the die-set buildup. Die-set mounting-hole locations are obtained from both platen drawings.

### *Pins*

Die-set thickness, pins, and bushings selected for the die specification in Table 11-3 are conservative and tend to be oversized (see Figures 11-21 and 11-22). Die-set platens of 2.50-in. (63.5-mm) thickness offer a solid foundation to support the punch-and-die cavities. Four 1.50-in. (38.1-mm) diameter pins and bushings provide precision and guidance for movement of the die set. The pins, which should always remain within the confines of the bushing, use a 4.50-in. (114.3-mm) press stroke. Pins can be as long as 11.00 in. (27.9 cm). The trimmed-part openings in the die plate can be flame-cut and later machine-finished to size.

### *Scrap Cutoff*

The scrap-trim cutoff location makes use of the trailing edge of the three cavities as a convenient blade rather than fabricating an additional component. This leaves an open area for the trimmed scrap to fall cleanly away from the die and avoids any uncut cavities from being crushed and possibly jamming the incoming indexed sheet.

The stripper plate clamps the trim scrap firmly against the die's blade edge. This assists in achieving a clean cut as well as stripping scrap away from the trim blade.

## Conclusion

Design of the trim punch and die does not take place in a vacuum, shielded from outside forces. Mold layout has a major influence on both the cost and ease of fabricating a die. The customer's thoughtful input and willingness to modify particular specifications when necessary can turn a complicated die project into a routine operation. Toolmaking facilities with modern precision equipment, skilled employees, and enlightened management ease the die-designer's task. Trim-press manufacturers are an essential part of the die team because they supply the equipment to help maintain the die in its original precision condition, provide

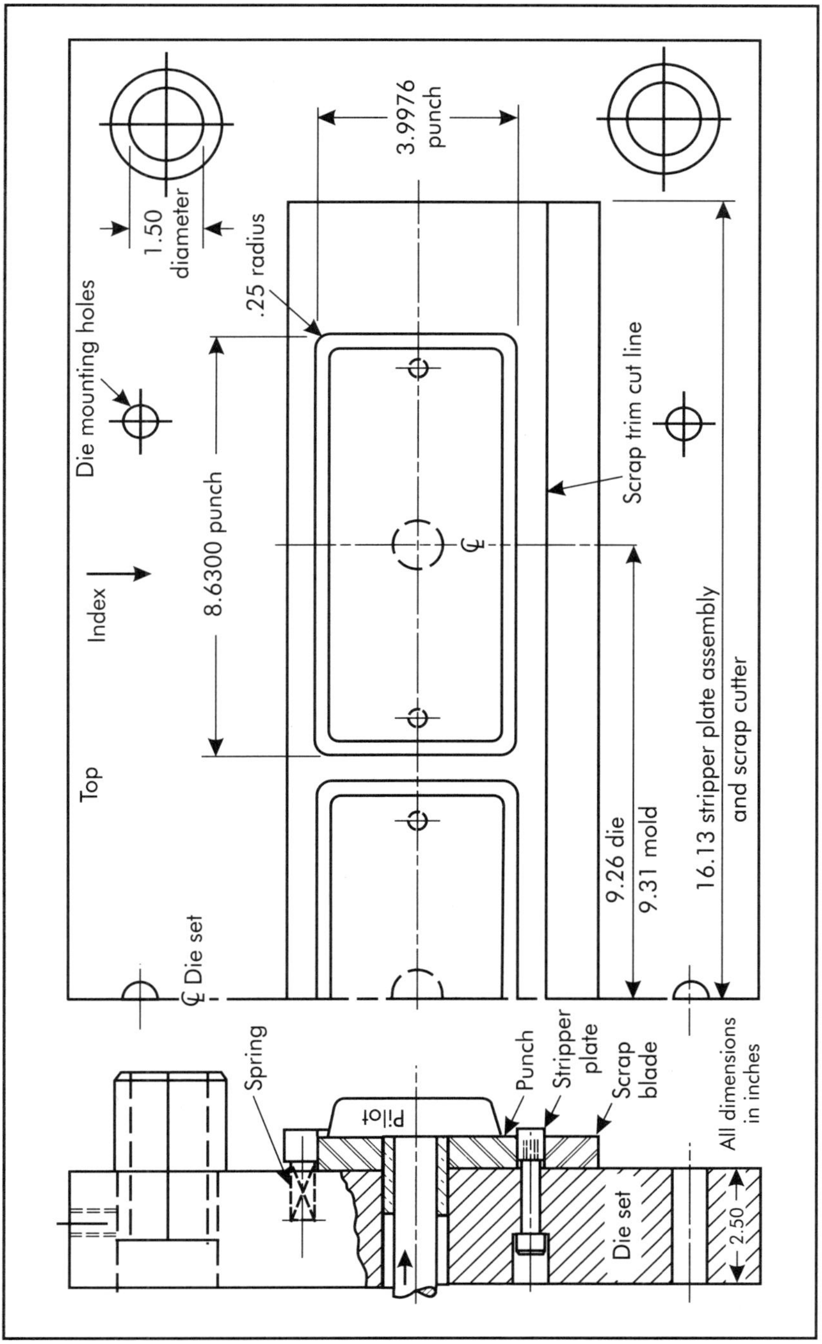


*Figure 11-21. Punch plate assembly.*

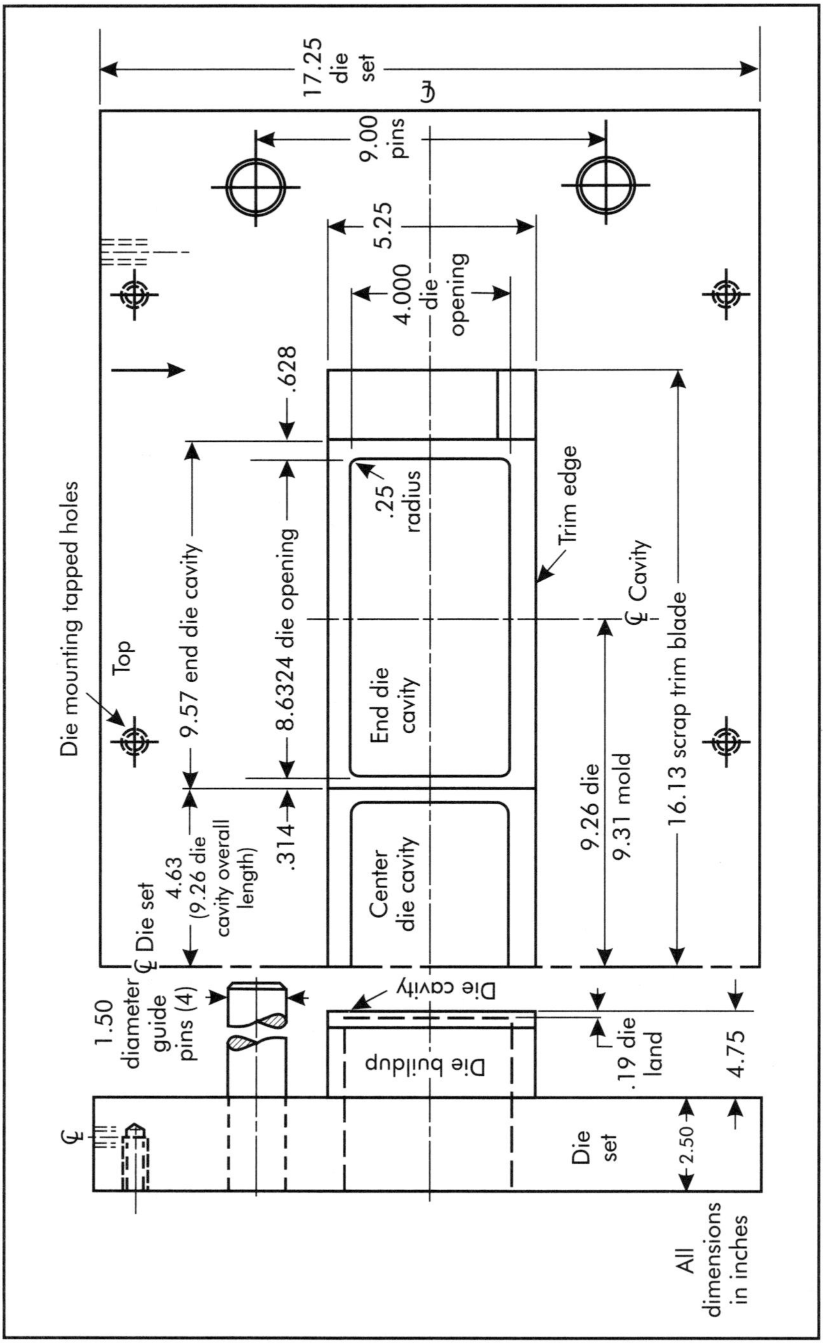


*Figure 11-22. Die plate assembly.*

easy access to mount the tooling, and maintain a safe environment for machine operators.

Finally, thermoforming shop personnel who have been properly instructed in maintenance and operation of the die ensure a long, productive life for the tool.

## References

Mitchell, Philip, ed. 1996. *Tool and Manufacturing Engineers Handbook*, Vol. 8. *Plastic Part Manufacturing*. Dearborn, MI: Society of Manufacturing Engineers.

Nee, John. 1998. *Fundamentals of Tool Design*, Fourth Edition. Dearborn, MI: Society of Manufacturing Engineers.

# Index

## D

## T

U



V



W



Z